Coloured Sheep

a colour genetics primer

About the authors

Irina Böhme was born on September 19th, 1973 in Rinteln. As a molecular biologist she worked with state-of-the-art DNA-sequencing methods. In what sometimes feels like a parralel universe, she breeds sheep. Since spending time in Ireland as an Au-pair she has been fascinated with sheep. The old nordic breeds being most dear to her heart. That fascination led to travels - long and short - to some of the breed's 'home turf': In order to learn about sheep, shepherds, their country and their history.

Sheep colour genetics combines her interests in sheep and science and her passion for making scientific knowledge accessible to non-scientists.

Saskia Dittgen was born on November 12th, 1968 in Heidelberg. She made her dream come true in 2009 when she bought a smalholding in Krahne/Brandenburg. She takes care of 2 Border Collies, several cats and chickens and about 40 sheep. Saskia breeds pedigree Shetland sheep since 2018. She is fascinated with colour genetics and she and Irina have been working through pedigrees of many generations of Skudde which led to the idea for a colour genetics primer.

Saskia works as a freelance editor and translator for Dutch and German.

Coloured Sheep

a colour genetics primer

Coloured Sheep: a colour genetics primer

Photo acknowledgements appear on pages 112-113

Irina Böhme
Mittelweg 39
01909 Grossharthau
Germany

books@driftwool.de

First edition

ISBN: 978-3-9820761-0-2

For Flora, Rosa *and* Violet

three white Skudde. Without them none of this would have happened.

Contents

Foreword

With great excitement I awaited this book. It was long overdue. We were forever entangled in endless discussions about the question: what colour is this sheep? And why?

Sound knowledge was needed: we needed scientific foundations! Sure, colour genetics follows biological rules, but what are they? Irina Böhme and Saskia Dittgen have put it on paper – finally!

The passion of both authors for the colour diversity of nordic sheep breeds combined with in-depth knowledge in the field of genetics and Irina's very individual style of writing give this book a lightness that will hopefully inspire many readers, like me.

Back in school genetics seemed awfully complicated and applying it to breeding sheep seemed impos-sible. Now it not only makes sense but turned into a fun game!

Sigrid Heilmann

I think this is an excellent book and I am sure that all breeders of coloured sheep would learn a lot from reading it. It is really well done and much more detailed and covers a wider range of topics than it appears to at first glance. I am very impressed with this work.

Roger Lundie

Is this book for you?

This book is for anyone who is interested in sheep colour genetics, who wondered about a black lamb out of two white parents, or a brown spotted one out of a white ram and a black ewe.

This book is for breeders who take delight in coloured sheep and who want to know where all this colour comes from, but it is also for those who want to avoid certain colours in their flock and want to learn how to do that.

This book is written for people without any background knowledge in heredity or genetics. It is a primer to help understand the basic principles of sheep colour genetics. To help open up a fascinating aspect of breeding sheep.

Introduction: coloured sheep

Nature is surprising, fascinating, and resourceful in creating diversity. That is also true when humans step in with planned breeding. Sometimes even more so when humans step in.

You probably have an image of an archtypical sheep in your head. But no single real sheep will match that 100%. There is lots of diversity. But we recognise those diverse animals as being sheep. In most cases.

Now think about colour: did you ever have a discussion about a shirt being turquoise, teal or light blue? Apparently everybody sees colours slightly differently. Most of us will have heard the cliché that women differentiate more colours than men. A discussion about 'teal versus blue' is probably not as common among men.

Colours and their distribution in the coat of a sheep (or other animal) and the genetics behind them cannot be assigned at will by a human observer. Genetics follows rules. We can deliberate over whether we call a particular colour 'grey' or 'blue' or 'red' or 'brown' or whatever we think describes it best. But the pigments that make up the colour and their distribution in the coat of a sheep won't be changed by the name we call it.

This books aims to help to SEE. And to explain the biological rules that colour genetics follows. It might also be helpful for finding a common language when discussing a sheep's colour. What one person might call a 'mottled red' could be 'brown' for the next person.

Left: Skudde lamb

'They all look the same!' and 'this is supposed to be the same colour?'

We'll look at differences that have a genetic cause (that we can then use for our breeding aims) and we'll look at similarities that can sometimes hide behind individuality. Not all differences in appearance have a (simple) genetic cause.

Some things we can see easily. Others are more of a puzzle. Sometimes we can only figure out what is inherited if we know a couple of related animals and follow this trait through the generations. This book will help with that, too.

It is not only about SEEING what is there but also about describing it in a concise way: we'll introduce some lingo for colour genetics. We won't need lots of complicated terms; mostly it is the common, everyday terms that cause misunderstandings in discussions about sheep colour. Using standards terms for colour genetics will help.

The most confusing is COLOUR itself! In this book we use colour to mean the overall impression we have of the colour of the whole sheep. 'It looks smudgy mottled blue' would be its colour.

This overall colour is made up of several components. One of them being the pigment. 'The main dye used for painting colour on a sheep' if you so wish. We call this the base colour.

Keeping sheep has a long tradition. They were (together with goats) the first livestock that humans domesticated. There are lots and lots of regional terms in use. We won't be talking about gimmers and hoggs.

We'll use **ram** for the (intact) male sheep and **ewe** for the female. We'll use **ewe lamb** and **ram lamb** for the young animals.

Breeding for colour?

For hundreds – no: thousands – of years sheep have primarily been used as a source of fibre. Colour is important if wool is being used. White wool can be dyed in lots of different shades. Coloured wool won't need to be dyed. There are many examples of folk costumes that have bodices, vests and coats made of naturally black wool.

Some breeds were mainly used for making pelts. A silvery grey pelt was often preferred.

Fashion and the market often influenced the breeder's decision and they still do today.

Those selling wool in large quantities to woolen mills need consistency. Large batches of very similar wool. In colour and quality.

A small scale farmer might take delight in a colorful diverse flock. Fibre artists are fascinated by all the hues that can be found in naturally coloured wool.

Colour is powerful. It is rather hard NOT to look at colour. It jumps at us and influences our impression more than anything else. (Except maybe horns. They are hard to overlook, too.)

But there is much more to a good sheep. Colour can be an added bonus. Knowing how colour genetics works can help us with our breeding decisions. Whether we aim for pure white or spotted or 'creamy cappuccino coloured'.

And it's great fun, too!

A simple model

Heredity – genetics – is pretty complicated. But predictable to a high degree with the knowledge of just a few basic rules.

One of the first who figured out that there are 'rules' to how traits are passed on to the next generation and the nature of these rules was the monk Gregor Mendel.

Mendel did breeding experiments with peas and came up with an idea to describe those rules that he observed. He published them in 1865, long before chromosomes and DNA were discovered.

Science has made a lot of progress since then and we know a lot more detail now. But Mendel's model is so good that it is still valid. Because it is a model based on clear observation. No more. No less. It is a system that helps us to visualise complex things in a simple way.

Lots of people remember Mendelian Rules from school. For some people they are very logical while others could never get their heads around them.

In this book we use a simple analogy to illustrate these rules:

Colour genetics as a game of cards

Left: brown Skudde ram

Colour genetics as a game of cards

A bit of luck, some skill and lots of fun: colour genetics is like a game of cards.

We need a bit of background knowledge. Or else this game would be like a slot-machine where we hope for 'three lemons' again and again.

The basic blueprint of a living being is coded in its genome, its **DNA**. DNA is organised in long stretches - **chromosomes**. Each chromosome has sections – **genes** – that are small, functional units. Being just a blueprint these genes need to be 'read' and translated. Just like a blueprint is read and turned into a house or a car or whatever the plan specifies. A chromosome could be likened to a recipe book with a gene being a specific recipe.

Sticking to colour genetics: the blueprint gives instructions to make certain proteins that carry out a range of 'jobs' in the body. "Go to this cell in this part of the body and incorporate this pigment". That is an image we can use to visualise what is going on.

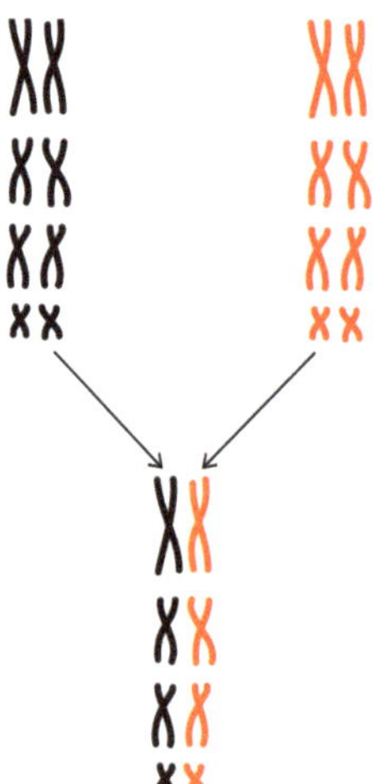

The offsapring inherits half the set of chomosomes from father and mother

A sheep has 54 chromosomes with lots of genes on them. They contain everything that is needed to make a sheep. We can picture this as a stack of 54 cards. Organised in pairs. 27 pairs – 27 different cards (different sets of instructions) in duplicate. One of each pair from mum, one from dad. Each chromosome – each card in a pair – has instructions for the same trait as the other card in the pair but the details can vary. Each sheep has this full set of cards. Giving half of them to the next generation. That is nature's way of trying new combinations. They might work better...

As a breeder we take part in this game. If we know the rules, we have a better chance of reaching our breeding goals.

'Half of the cards' can't just be passed on randomly. A sheep could end up with 10 instructions for legs and no instructions for ears. They can be different versions but they are for the same trait. Those variants that exist for a certain gene are called **alleles**. We'll use this term a couple of times throughout the book.

As an example: chromosome 10 (or card No 10) has – among lots of other stuff – the information for 'horns'. The possible variants (as far as we know) are: 'horns yes', 'horns no', 'horns on male animals only'.

Note

DNA: The material that contains the code for genetic characterisits in all life forms
Chromosome: Stretches of DNA that are bundled up
Gene: One unit of information on a chromosome
Allele: One variant of a gene

!

So we picture the 54 chromosomes as cards; 27 pairs of cards. Imagine them as 27 stacks with two cards each. The next generation gets one card of each stack – one from each parent. So the lamb will end up with 54 cards altogether. Two on each stack. Nothing missing, nothing extra.

We have all the instructions for each part of the sheep. All in duplicate. Instructions that are specified on the same card can only be passed on (inherited) together. (Which is not entirely true – nature has found some clever ways to mix 'bits of cards' but that is not the norm and we can ignore it for now.)

The colour game

Sounds complicated? 27 pairs of cards on 27 stacks... Which of the cards is it that says 'colour'? The good news: we don't need all 27 pairs for colour genetics. The bad news: ONE isn't enough either. There is no single card that says 'make this sheep mottled reddish brown'.
More information is needed. What pigment? Where? How much?

The overall colour of a sheep is basically made up of three components:

- base colour
- pattern
- spotting

Three genes on three different stacks (chromosomes). Each with different variants (alleles). All variants of one stack can be freely combined with those on any other stack. There is no biological barrier that prevents a certain base colour from being combined with a certain pattern.

This is where breeding for colour turns into a fascinating and exciting game.

We need knowledge, a good eye, clever strategies, and some luck.

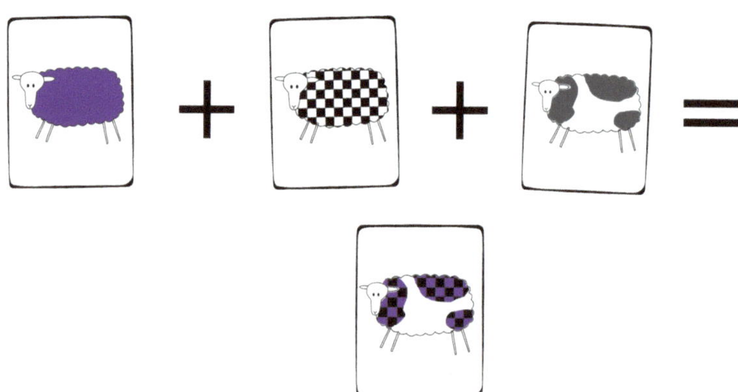

base colour purple + pattern checkered + spotting = colourful

The rules

- Three stacks with two cards each per animal. One from its mother, one from its father.
- Of each of those three stacks ONE card will be passed on. It doesn't matter if a card was inherited from father or mother. It is pure chance which one is chosen.
- The three stacks are played independently. No matter what was played on the other stacks.
- On each stack one card 'trumps' the other. That is the version of the trait that we can see.

The rules are rather simple. But the combinations of cards in each stack and the combination with cards on the other stacks makes for an amazing amount of variety!

If we look at a 'coloured sheep' we can focus on those stacks independently. Divide the overall impression into components. We will see similarities and differences. This sheep – and that over there – both have the base colour black. But one is also spotted!

Shetland lambs. Both have the same pattern – but on different base colours. left: brown right: black

The colour game: warm up

Learning a new game just in our imagination is a challenge. Therefore a set of cards is included in the back of the book so you can play along.

Before we look at the cards in detail: let's play a little warm up game!

We have our three stacks: base colour, pattern, spotting.

Each stack has two cards. They can be identical within a stack or different.

Cards for pattern come in a huge variety of possibilities. Lots of alleles exist. The other stacks are way simpler because only two versions exist for each of them.

If we have two identical cards on a stack, we call this animal **homozygous** for this trait. If there are two different cards, we call it **heterozygous**.

Left: Skudde

Base colour

Base colour is the version of pigment that is used. Actually the version of a certain type of pigment: **eumelanin.**

Eumelanin comes in two versions: **black** and **brown**. That's it. With this stack of cards we play for the version of eumelanin that is produced. IF and WHERE it is incorporated (and shows) is something we'll gamble about later.

> **!**
> ### Note
> ***Every* sheep has two cards for base colour**

So we have the two possible cards BLACK and BROWN. BLACK is the dominant allele. BLACK trumps BROWN. If there is just one of the cards saying BLACK, the animal will be black. If the second card is BROWN, we won't see it. We'll see black. Whether the other card is BLACK or BROWN.

Therefore a brown animal needs to have two cards saying BROWN. A BLACK card would trump the BROWN card.

Skudde rams: brown and black

Let's play with just this stack: our first move in the game is just about base colour.

Let's say we have a brown animal. We know both its cards. Both must be brown. For a black animal we only know one card. One is black. The other one can be either black or brown. No matter what it says on the second card: the animal looks black. That is called **phenotype**: what we see.

In a brown animal the phenotype is the same as the **genotype** (what is on the cards). Genotype describes what is coded on the genes. Whether we can see it or not.

If we ignore the other stacks / the other genes for now, these are the possibilities for base colour:

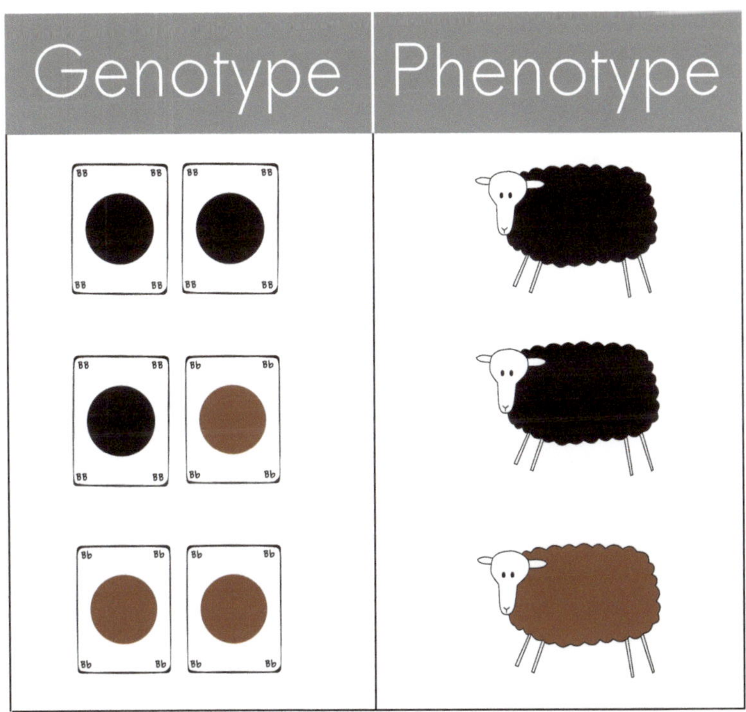

Brown ram and heterozygous black ewe:

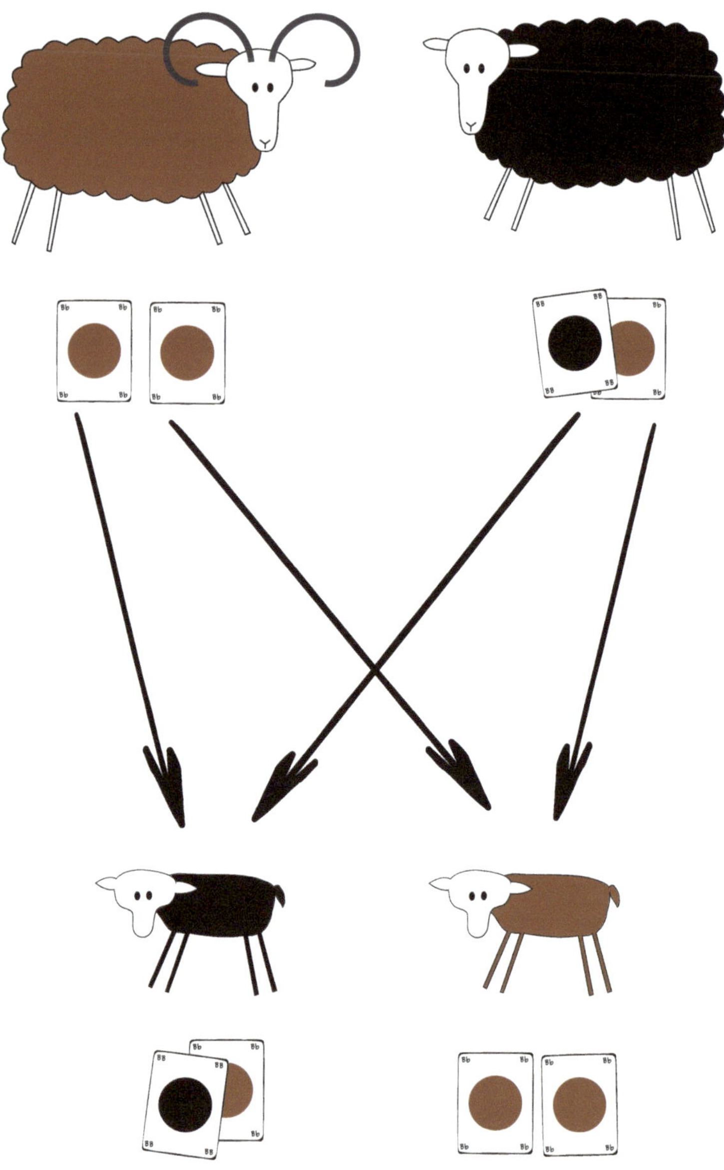

A brown ram and a heterozygous black ewe can have black lambs as well as brown lambs.

Let's play for base colour:

We have a brown animal and a black one. Each giving one of its cards to the offspring. Of course they can only pass on what they have. They can't play with cards they don't have. The brown animal can only pass on BROWN. Nothing else is in the stack. The black animal can pass on the BLACK card. It must have at least one BLACK one or else it wouldn't be black. If it passes on a BLACK card, the lamb will have a BLACK card and a BROWN card. BLACK trumps BROWN – the lamb will be black.

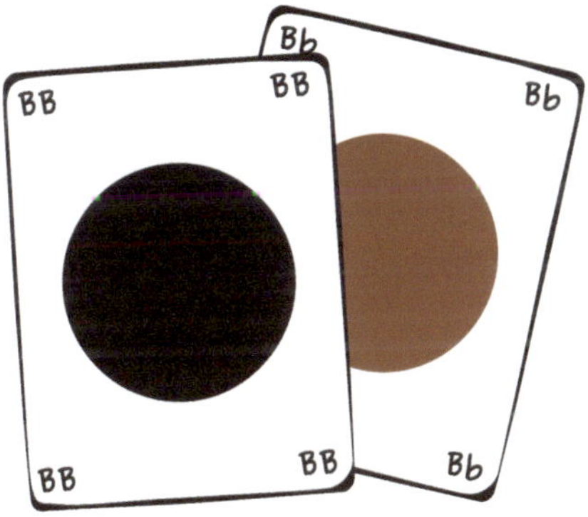

If the black parent has another BLACK card for its second card, the lamb will of course be black too. But the black animal could have BROWN for a second card. If this card is played, the lamb will have two BROWN cards. It will be brown.

BB BB BB BB

Bb Bb Bb Bb

With this stack – our first move in the game – we played for the base colour that is available: black or brown.

cross breed lambs, black and brown

Skudde with lamb, white

Icelandic sheep, solid

Skudde, grey

Pattern

The gene for pattern determines where this base colour is used. In our first game we'll just play with three cards. Three patterns. There are more but we'll keep it simple for now. It works just the same with more variants.
So for now we use:

White: White is not a 'colour' – it is a pattern. The instructions read: don't use base colour anywhere!

Solid: The instructions for solid read: use that base colour everywhere!

Grey: This is a tricky one. The instructions read: use a bit of the base colour here, a little more there, none at this bit, and loads over there!

More precisely: very little around the mouth and nose, a lot on the rest of the head and the legs; some hairs on the body have lots of eumelanin but others have very little.

Not in patches but one hair very dark, the next very light. The overall impression of a mix of black and (nearly) white hairs is grey. BUT: grey is not a good name for a pattern that works on a base colour that can be either black or brown! The animal would probably be called fawn if the base colour was brown.

Think of AGOUTI GREY as a pattern and ignore that you have an image of 'grey'. In this case it is just an instruction for where and how much of the available base colour is used: a pattern.

These are the cards we are playing with for now:

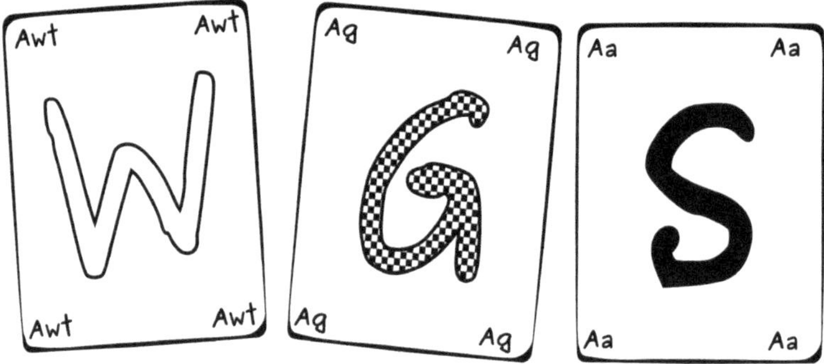

W for WHITE, G for GREY and S for SOLID

There are three of them but every individual animal will have only two of them and will pass on only one to the lamb. The rules for patterns are:

- WHITE trumps GREY trumps SOLID. If there is one card for WHITE the animal will be white. Regardless of the second card this animal has.
- GREY trumps SOLID. If there is one card for GREY (and none for WHITE) the animal will be grey. (Remember: **Pattern GREY**. In combination with a brown base colour it will be fawn)
- The animal will only be solid if both cards are SOLID.

Phenotype | Genotype

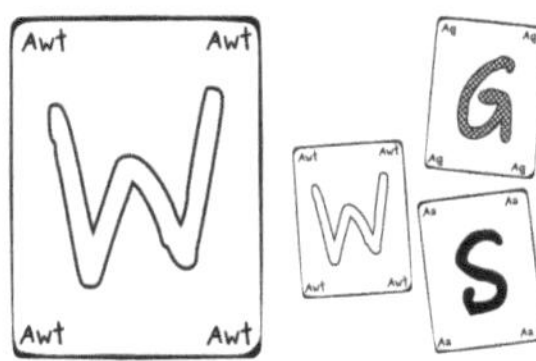

If an animal is white (if the phenotype is white), it has one card for white.
The second card can also be WHITE. *Or* GREY. *Or* SOLID.

A grey (or fawn) animal has a card for GREY. *The second card can also be* GREY. *Or* SOLID. *It cannot be white. The animal would be white if it had a card saying* WHITE.

With a solid animal we know both cards. Both must be SOLID. *The phenotype is the same as the genotype. We know what card this animal is going to play. It has two cards saying* SOLID *and it it will pass one of them on.*

Spotting

UN-SPOTTED trumps SPOTTED. A spotted sheep has two cards that say SPOTTED.

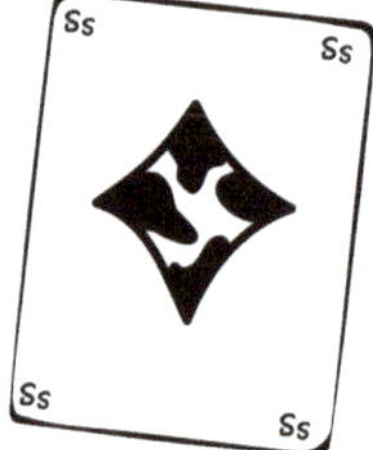

A sheep that does not show spotting has one card saying UN-SPOTTED and another one that can be either SPOTTED or UN-SPOTTED.

Spotted means that parts of the animal are white and others are coloured. There are bits where no pigment is incorporated. It is as if parts of the sheep are not painted or coloured in. Spotting ranges from a single tiny white area to nearly white animals. There can be large blobs of colour or small ones.

Spotting is our third stack of cards. The last one. This stack is also played independently of the others. Whatever cards are on the other ones – they will be combined with whatever is on this stack.

Here is another great opportunity for confusion: You could talk about a 'solid spotted animal'. Surely a black and white or brown and white animal is not solid? Well, it isn't if you think about solid as being the same colour from nose to tail. But it is solid if you use that term in the convention of sheep colour genetics. Meaning 'those bits that do get painted will be painted in a solid colour that is defined by the gene for base colour'.

Saying 'solid spotted' just means that the non-white bits have the pattern solid.

It helps to think of the three stacks. Spotting defines where colour is incorporated. The whole sheep or just bits. Pattern defines how those areas with colour are coloured. The variant of eumelanin is defined by base colour.

When talking about 'white' in a spotted sheep we are talking about white patches. Not about the pattern WHITE. A black and white sheep is not 'partly pattern WHITE, partly pattern SOLID'. It is homozygous SOLID on the pattern locus (having two cards for SOLID – anything else would trump SOLID and make it invisible for us). It is also homozygous for SPOTTED – having two cards for SPOTTED. (UN-SPOTTED trumps SPOTTED – spotting is only visible if both cards are SPOTTED.)

If we can see a version (on either stack) that is trumped by the other possibilities, the animal must have two cards saying the same thing. It must be homozygous for the recessive allele – the version that is trumped by the others.

Skudde lamb, black and white spotted

Shetland sheep (Foula)

Skudde

Phenotype

Genotype

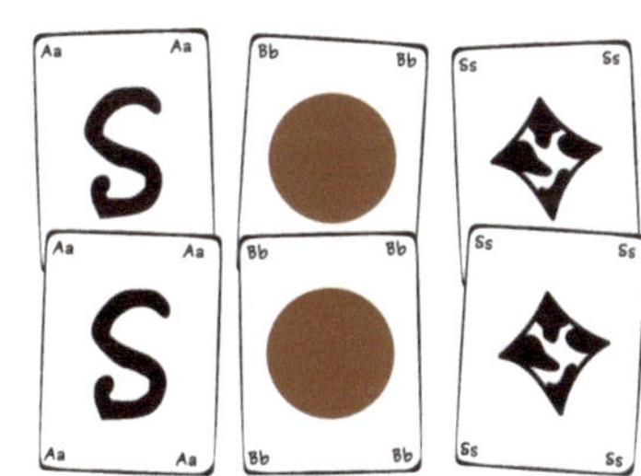

A solid brown spotted animal doesn't have any unknown cards. We know the hand it has and the possible cards it can play. It must have two cards for SOLID, *two for* BROWN *and two for* SPOTTED.

It is quite the opposite for a white sheep. Only one card is known. It has one card for WHITE. *It still has six cards – two on every stack – like any other sheep.*

Two for base colour BLACK *or* BROWN *and two for either* SPOTTED *or* UN-SPOTTED. *But we can't see what they are. Pattern* WHITE *gives the clear instruction NOT to incorporate eumelanin (the pigment of the base colour) anywhere. A white animal can also be spotted. But white spots on white are a little hard to see...*

Colour game for beginners

We know the rules – we did a little warm up: Let's play! We'll just use the cards we already know.

- **Base colour:** BLACK and BROWN
- **Pattern:** WHITE, GREY, SOLID
- **Spotting:** SPOTTED und UN-SPOTTED

We have a brown spotted ram and a white ewe. We know the ram's hand – the cards he has and could play.

BROWN, SOLID, SPOTTED. Two of each.

The ewe can have all sorts of surprising cards in this scenario. We just know that she has one card for white. If she puts that card on the lamb's stack for pattern, the lamb will be white. It looks the same as the ewe whose cards we don't know.

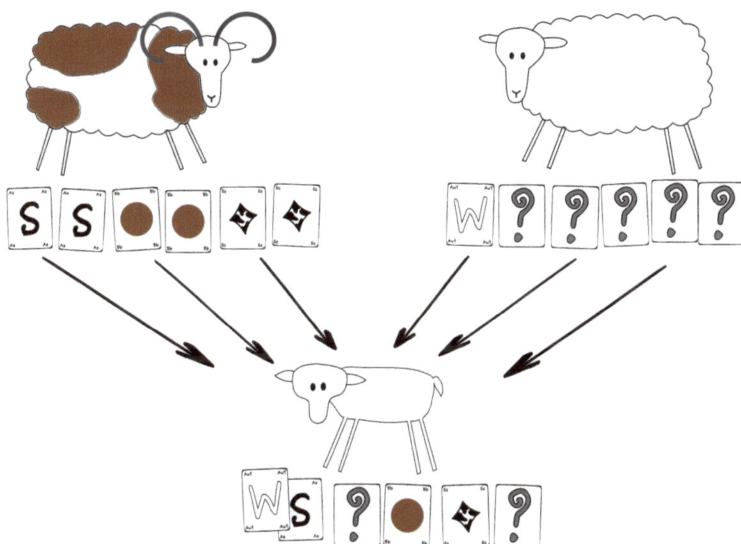

If one of the parents plays a white card all the other cards that the lamb has will be invisible to us

Left: Skudde lamb

But in the case of the lamb we know more! We know that the ram passed on one card for brown, one for spotted, and one for solid. That's all he has and he must pass one of each stack on. And we know both the lamb's cards for pattern: white and solid.

Let us play the next round of the game. Let's say our lamb is a ewe lamb and we breed it to a ram the same colour as the first ram. This new ram can also only play BROWN, SOLID, SPOTTED. If our (now grown-up) ewe lamb plays its WHITE card on the pattern stack, we know just as much as before: we know what the ram added but we don't know what the ewe has and what she played. Except for the WHITE. This new white lamb has one card for SOLID, BROWN and SPOTTED and one for WHITE. That leaves us with one unknown for base colour and one unknown for spotting. The unknowns that might have come from the original white ewe. We can't know if those got played or the known cards that came from the original ram.

What we can discover from this example: white can hide practically everything. For generations. One white card played: white animal. No matter how often an animal is bred: it can always play the same card(s). And the next generation can do the same. We might never know the other cards in the game. It is a matter of probability.

Another scenario 'brown spotted ram and white ewe':

The ram can again only play BROWN, SOLID, SPOTTED. But the ewe now plays another card: GREY. We couldn't see that she had this card hidden. Her WHITE was masking it – hiding it from our eyes.

So now the lamb has SOLID and GREY on its pattern stack. GREY trumps SOLID so it will be grey. We don't see the SOLID but we know it is there because we know the ram and his cards. If he didn't have two cards for SOLID he wouldn't be solid. And since he only has SOLID he must have played it – passed it on to the lamb.

Now we not only know both pattern cards for the lamb. We also know that the ewe's second pattern card is GREY. One WHITE, one GREY. This ewe will give us white or grey lambs.

Even with a solid ram we will only get white or grey lambs. The ewe must play one of her cards. WHITE and GREY both trump SOLID.

Back to our lamb:

The visible **pattern** is GREY but we know about the SOLID.

Base colour: Now that we have a card for pattern that actually uses base colour, we can see the variant of base colour. The ram played a BROWN card. If the ewe also played a BROWN card, the lamb will end up brown. If the ewe played a BLACK card, the lamb will end up black. BLACK trumps BROWN. Either way we know both cards for the lamb and one for the ewe – the one she played.

If the lamb is black, it is a black based grey. It will be born black and over time lighten to grey. If it is brown, it will be born brown and over time lighten to fawn.

Third stack: **spotting**

Our ram played SPOTTED. If our lamb is not spotted, the ewe must have played an UN-SPOTTED card. If the lamb is spotted, she must have passed on a SPOTTED card. Again we now know both cards for the lamb and one for the ewe.

We now know a lot more about the colour genetics of our animals!

But it is still about chances and probabilities. Hundreds of generations of white sheep can still pass on another pattern card that will remain invisible. If all the sheep in our flock have at least one WHITE card and some have two WHITE cards, then chances are high that white is all that we'll ever see. Which doesn't mean that white is all that exists in our flock! We can only know about other cards if they are actually played. And end up in a combination that allows us to see them.

Two white sheep out of generations of white sheep can both have a card for SOLID. And play it. Only then do we know about their existence. And only then can we see base colour. We might have the black sheep of the family. Or the brown sheep if both played BROWN.

Skudde lamb

Note

Every sheep is either black or brown - including white sheep!
If there is a dominant allele (a card that trumps another), we can't see what else is there (what is on the card that is being trumped).

Time to play a game on your own:

We know the genotype of the ewe. She has a card each for BROWN and BLACK. On the pattern stack she has two cards for SOLID and for spotting she has SPOTTED and UNSPOTTED.

The ram is grey spotted. Pattern grey on base colour black.

Our ewe has twin lambs. One is not spotted, shows pattern grey, and base colour black. The other lamb is brown spotted (solid brown spotted that is).

What is the genotype of those lambs?

What do we now know about the genotype of the ram?
(See page 137 to find out if you're right)

Villsau (old norwegian breed), grey spotted ram

The influence of fleece type

How we see colour (the overall colour made up of base colour, pattern, and spotting) is not only influenced by the pigments and their distribution but also by many other factors. One of them is the type of fleece.

Very fine wool fibres tend to incorporate pigment less intensely and they also tend to bleach more easily in sunlight. Therefore the same colour genetics will present differently in sheep with different fleeces: Hair sheep often have very clear and bright colours. Fine wool sheep rarely have a deep 'midnight black' or dark chocolate brown. Double coated sheep show both. They have a hairy top coat (guard hair) and fine wool fibres.

Most sheep have thick short hair fibres on face and legs. Some breeds also have thick short fibres (kemp) in their fleece.

Fibre types of a primitive breed (Skudde): Guard hair, wool fibres, kemp.

Left: Icelandic sheep, double coated

These differences change the appearance of colour. A black and white friesian might have dark black on its face and legs (where it isn't white through spotting) but the wool could look very much like brown. A black and white spotted Icelandic has 'hair' not only on its face but also on its body. These guard hairs will often be darker than the wool fibres and they also protect the shorter wool fibres from bleaching in the sun.

Since the body with its wool is the biggest part of a sheep, we tend to look there and choose a colour name based on that impression.

Note

If we want to know the base colour, we need to look at a sheep's face and legs.

This is where 'base colour' can help clarify things. The overall impression of a sheep might be 'brown' but looking at its black face we see that the base colour is black.

Looks like a brown sheep at first glance. But after shearing we can see that the black wool faded to brown on the outside of the fleece.

Colour game for advanced players

We looked at how colour genetics works in principle when pictured as a game of cards: the stacks (the genes on their chromosomes) with different versions of cards (alleles) on each stack. Some cards trump the others on each stack. We'll now look at our three stacks in a little more detail.

Base colour

We'll keep playing with just two cards. That is all there is. We use 'base colour' for versions of eumelanin (and 'colour' for the overall colour we see in a sheep – which is made up of the effects of all three stacks (genes) combined plus environmental effects like bleaching in sunlight and even nutrition).

There are only two versions of the pigment eumelanin: black and brown.

Eumelanin means 'true' or 'real' melanin. Depending on what is coded on the gene for base colour one of two versions is incorporated. There is variation in each one – brown for example ranging from a reddish chestnut to a dark coffee colour.

It can get confusing if we use words differently. Especially if we don't realise that someone else is talking about a different thing.

There is a convention on what to call the alleles and also a handy set of abbreviations to take notes and exchange them with others.

For base colour these are:

BB for BLACK and **Bb** for BROWN

The first letter denotes the gene. Or rather its **locus** which is the term used for a location on a chromosome. The place. So it is pretty much the same as 'gene' in our context. We are talking about genes and those are on certain loci (which is the plural of locus). Other loci can have stretches of DNA that are not genes – so talking about a locus is broader.

B denotes the *Brown-Locus*. It could have been the *Black-Locus* just the same but it got called the *Brown-Locus*.

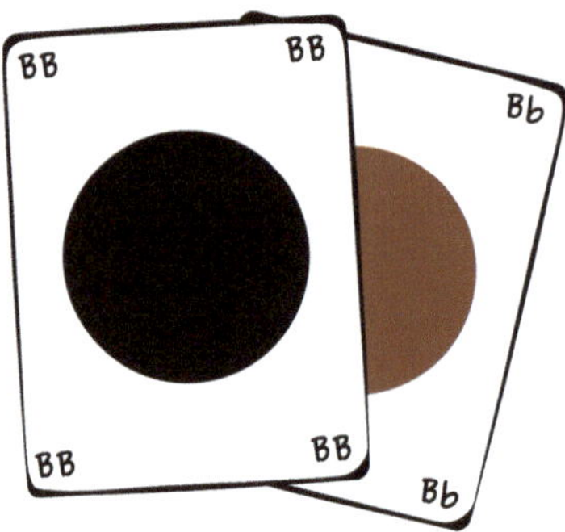

The second letter denotes the variant. The **allele**. A capital letter is used for the **dominant** allele (the one that trumps) and the lower case letter for the **recessive** (the one that gets trumped).

In our visualisation with cards we put the card that trumps on top of the other card that is being trumped.

Additional Information

The gene itself that is located on *Brown-Locus* is called TYRP1 (Tyrosinase related protein 1) and is situated on chromosome 2.

Patterns - *Agouti-Locus*

So far we have played our game with just three cards for pattern. But there are more. The three we already know are:

- **White**
- **Grey**
- **Solid**

These patterns (and the others that we will introduce shortly) are coded on chromosome 13. This gene has the information for ASIP - Agouti Signaling Protein. The locus - the place where this gene is on its chromosome - is called *Agouti-Locus*.

In the long history of sheep an astonishing variety evolved on this locus. A lot of this variety was (and is) preserved in old breeds - the sheep of remote islands, fjords, and valleys.

Some breeds just have a single colour nowadays. Others have a huge palette. Among the most colourful are Icelandic sheep and Shetland sheep.

Colourful flock in Norway

But even in the ones with just one colour we do get surprises every now and then. As we have seen it is white first of all that can 'hide' things over generations.

If we try to figure out the colour genetics of a sheep, it helps to look at its breed (or breeds in its ancestry if there are more than one) and the patterns that are normally found in this breed. Always keeping in mind that we might be seeing something else. A pattern that is very rare (in this breed), a pattern that people didn't notice before, or a pattern that was mistakenly thought to be something else.

But it helps to start with what is common and known. If that fits what we see: fine! If there are doubts: Look further!

The names used for patterns by shepherds can differ a lot between breeds and between regions. What is called katmoget in a Shetland sheep is called golsótt in Icelandics (in Iceland) and dassenkopp in Drenthe heath sheep in the Netherlands.

When talking about phenotype it is usually best to use what is commonly used for your breed and in your area. Especially if you breed registered sheep. Of course for yourself you can use whatever sounds good to you, but if you advertise sheep for sale for example, it is better to stick to what others know and expect.

Agouti Signaling Protein

Agouti Signaling Protein is what is coded for at the *Agouti-Locus*. So the construction plan – the blueprint – gives instructions to build this protein. As the name suggests, the Agouti Signaling Protein then has the job of signaling. It signals if eumelanin is going to be used or not, in whatever version is coded for on the *Brown-Locus*. Basically eumelanin is used as long as ASIP hasn't got other plans. Those other plans can be 'use just a little' or 'use none at all' or 'use another type of melanin'. That other type is called **pheomelanin**.

Remember that the *Brown-Locus* codes for eumelanin and eumelanin comes in black or brown. ASIP now controls if and how much of that pigment is used (whichever version) and it also has this additional pigment at hand. Pheomelanin is a yellowish to reddish brown.

Agouti Signaling Protein can use eumelanin or pheomelanin and it can use a little or lots or none. Signalling different things on different parts of the body and even at different points in time. The latter will result in banded hairs where eumelanin or pheomelanin or nothing is incorporated in turn.

In the wild ancestors of our domesticated sheep Agouti Signaling Protein has a lot to do: If we look at a moufflon we see that there is very little pigment on the belly, other parts have a lot of eumelanin, and other parts again have strong pheomelanin making for a very colourful animal.

Moufflon: White (no pigment), black (eumelanin), reddish-brown (phaeomelanin)

Let's look at some patterns in sheep:

White

We already know WHITE. No eumelanin anywhere. That is what this version of ASIP signals in this pattern allele. We can have pheomelanin, though! That can range from 'freckles' on the face to foxy red blobs on the body to a foxy red sheep. In many cases it is clearly visible in a young lamb but lightens to white (or near white) in an adult.

Shorthand for this allele is **Awt** (Agouti white tan). Tan being the reddish hue caused by pheomelanin that CAN be present.

White trumps everything else. In a list of dominance of agouti alleles it is on the top.

Skudde, white with freckles

Grey

Another one we already know. The adult animal has a light colour around its mouth and eyes. The wool is grey if the base colour is black and fawn if the base colour is brown. Just above their feet there is a ring of light coloured hair.

Grey is not really a good name for a pattern that works on a base colour that can be either black or brown! But common usage and tradition has it that we call the pattern itself after just one of the possible phenotypes it causes.

There are other patterns that cause the wool of a sheep to be grey or fawn and by just choosing a word by the overall impression those sheep might be called grey or fawn as well.

For the precise, unambiguous notation of the genotype **Ag** is used. (Agouti grey)

Skudde, grey

A few white hairs around the mouth give it away: He will be grey

When a few weeks old the white mouth and eye rings can be seen clearly - pattern grey on base colour brown

Left: Ag on base colour brown. Right: Ag on base colour black

As if the linguistic confusion wasn't enough, agouti grey animals are different to animals with other agouti patterns: Agouti grey lambs are born with just their base colour. They look very much like a solid black or brown animal. Over time they get lighter and we can clearly see the typical light coloured mouth, rings around the eyes and light coloured hair above their feet.

Very often newborn lambs give us a hint that they are not solid and will change colour soon: a few white hairs on their lips (sugar lips) and in their ears. Lambs with the agouti pattern grey will have pink tongues.

Animals with the pattern agouti grey often change colour through the seasons, too. Getting much darker and lighter.

Note

Agouti grey lambs are born black or brown

Drenthe Heath sheep - base colour black (in this breed the pattern is called dassenkop)

Shetlands sheep - base colour black (in this breed the pattern is called katmoget), solid brown sheep in the background

Cameroon sheep - base colour black

Badgerface

This pattern is common in Icelandics, Shetland sheep, Barbados Blackbelly, Finnsheep and others. In Badgerface Welsh Mountain sheep it is the only accepted pattern.

These animals have dark stripes on their faces - reminiscent of a badger. Their belly is dark. Or rather their whole underside is dark. From chin to the back of their hindlegs.

Lambs will often show strong pheomelanin - making the main part of the sheep look very red or brownish. Even if the base colour is black. It's the pheomelanin (tan) that we notice. But the dark bits on face and underside will tell us whether the base colour is brown or black.

Wool sheep will often lose this foxy red appearance as they age whereas hair sheep often keep the reddish tint .

The abbreviation is **Ab** (Agouti badgerface).

Black and Tan

Black and tan is another pattern that is often found in Icelandics and Shetland sheep.

Black and tan looks a bit like the inversion of badgerface. The upper side of the animal is dark (eumelanin) and the underside is light coloured. Chin, front of neck, belly and back of hindlegs are very light coloured to reddish (pheomelanin/tan). They have a characteristic light coloured stripe over their eyes.

The abbreviation is **At** (Agouti black and tan)

Shetland sheep, At - base colour brown (in this breed this pattern is called gulmoget)

English Blue

This pattern is easily overlooked. At first glance the animal looks very much like a grey. It's the head and legs we need to look at to see the pattern. Instead of the white mouth of an Ag sheep these animals just have a moustache. There are no rings around the eyes but very distinct 'tear drops'. And the body also shows a difference: Neck and shoulders are often very dark and the lighter colour looks as if it was applied with a sweeping stroke of a paintbrush.

The tear drops are visible at birth though sometimes they are not very clear so we need to take a close look. The grey or fawn of the body develops as the lamb matures.

The abbreviation for english blue is **Aeb**
(Agouti english blue)

Skudde, English Blue – base colour black

Light Badgerface

This is another pattern that is not very easy to recognise. It looks very much like badgerface. Light on top, dark on underside, badger stripes on the face. We can think of it as a version of badgerface.

Light badgerface and badgerface are different patterns that are just a little tricky to distinguish: Light badgerface looks a bit more 'smudgy' - less clear - than badgerface. Some of the distinguishing features of light badgerface are the light chin and light coloured hair around navel and purse. Also the front legs often show tan on the front whereas this part is dark on a badgerface.

This pattern is rather rare. But it could be more common than we think in some breeds due to its being easily confused with badgerface. It is known to exist in Shetland sheep and Skudde. Both are breeds of the northern short tailed group of sheep. It probably also exists in other sheep of this group.

The abbreviation is **Albf** (Agouti light badgerface)

Skudde, Light Badgerface - base colour black

Solid

We already know solid. In this pattern the Agouti Signaling Protein does nothing. More precisely: There is no Agouti Signaling Protein! The switch that turns the gene on is broken and therefore the blueprint can't be read. Without a legible blueprint no protein is made. There is no mechanism that prevents eumelanin from being used. It is 'eumelanin everywhere!' Whatever version is coded for at the locus for base colour.

The abbreviation is **Aa**

A lower case a is used here because there is no pattern that could be abbreviated. So the letter of the *Agouti-Locus* is used in lower case.

'Solid' is also called 'non-agouti' which is more in line with what is happening behind the scenes but less des criptive in terms of what the animal looks like (if the animal is homozygous – has two cards for Aa).

All the other patterns trump SOLID.

Skudde, solid -base colour black

Simply speaking we can think of the pattern alleles as a list sorted by dominance. Solid (or non-agouti) is lowest on the list of patterns, white is on top.

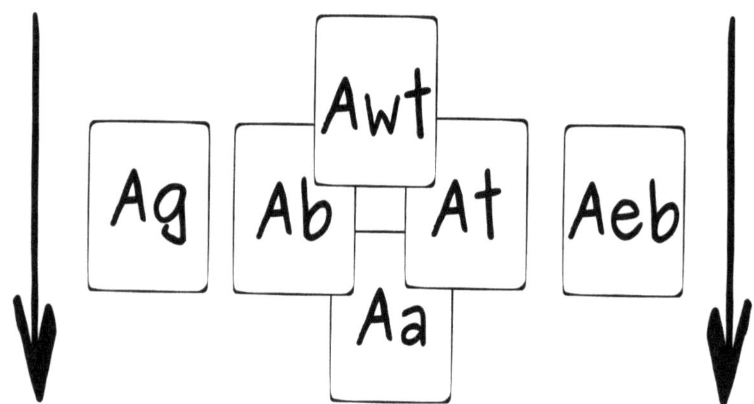

If an animal has one card for white, it will be white. The patterns Ag, Ab, At, Aeb, Albf (and some more that we haven't talked about) are dominant to solid.

An animal with one card for either Ag, Ab, Aeb, or Albf and another one for Aa will not show the Aa. We just see the other pattern.

Put differently: light wins. The card for white calls for 'light' everywhere. (More precisely: pheomelanin in varying amounts everywhere and no eumelanin.) So that is what we see. No matter what the other card says. If there is no WHITE card involved and two different cards are combined, the 'light bits' will also win.

When two patterns are combined, there'll be two versions of ASIP or twice as much of the same version. The Agouti Signaling Protein doesn't incorporate eumelanin. It just uses pheomelanin in differing amounts: lots of it – resulting in very intense tan; or very little to none – resulting in very light colour or white.

Each version of ASIP controls where and how much pheomelanin is used. Only where ASIP doesn't have its own plans is eumelanin used.

Therefore a mix of two patterns will result in both ASIP variants doing their bit of painting and only where none of the two has other plans will we see pheomelanin.

Let's say we have badgerface, which calls for 'light on top', and black and tan, which calls for 'light on the underside': we will only see eumelanin where none of the patterns calls for 'light' (regardless of whether that is very little pheomelanin or lots of it). In this case we might see a thin dark stripe on the side of the animal where 'top and bottom' instructions meet, on top of the nose and a trace of the badger stripes.

Having two versions of the same pattern as opposed to one version in combination with Aa very often results in a slightly different phenotype. There is twice as much ASIP that incorporates pheomelanin and prevents eumelanin.

If we have an animal that is a puzzle and could be all sorts of pattern combinations by the look of it, we can breed it to a solid sheep. A solid sheep has two cards for Aa. Or

eyebrows of a badgerface, white mouth of an Agouti grey. Ag/Ab, base colour black

else it wouldn't look solid. So it can only pass on solid. Our mystery sheep will pass on one of the two cards it has. Thus we will end up with one of them in combination with Aa. The pattern will be visible. If we are lucky and more than one lamb is being born, we might even see both patterns. Each in combination with Aa and each easier to recognise.

One thing to keep in mind: those patterns are just umbrella terms. There might be lots of distinct versions that are so minute in their differences that we can't discern them. Also other factors influence the appearance. Like wool type or age greying or diet and many, many more. Two animals with the same genotype for agouti might look different. But they are similar enough that we can use 'badgerface' for both.

Here, too we use a model. We build categories that help us to sort the fascinating diversity into manageable chunks of information.

In the case of light badgerface and badgerface there are differences that are distinct enough to put them in two categories. But they are not as easy to see as the differences between badgerface and grey for example.

There are many more agouti alleles than the ones we looked at. They were identified as being versions of agouti (as opposed to other influences on the appearance of individual sheep) mostly through test breeding and pedigree analysis. For example there are quite a few that are called 'blue' (with an adjective to distinguish between them like 'dark blue'). And there are versions of grey and of other patterns. Some of these versions are important in specific breeds but were never identified in others. Going into all of them in detail goes way beyond the scope of a sheep genetics primer but it is valuable information to keep in mind.

Spotting

With our third stack we play for spotted or un-spotted. Remember: UN-SPOTTED trumps SPOTTED. Both cards need to be SPOTTED for an animal to be spotted.

The abbreviations for the genotypes at *Spotting-Locus* are **SS** and **Ss**.

Spotting having its own stack means that it is played independently of the others. No matter what is in the stacks for base colour and pattern: the animal can be spotted or un-spotted.

The cards in the other stacks define what the coloured spots look like: brown or black; solid, badgerface, grey, ... or white! In that case the coloured spots are not coloured but white. So we can't see a difference.

Depending on where those coloured spots are and how big they are we can have a hard time figuring out what is happening in the other stacks. If for example head and belly are mainly white we might not be able to distinguish a spotted grey from a spotted badgerface.

Where exactly the white and coloured parts are on a spotted animal is also influenced by genetics. But apparently not a hundred percent. The type of spotting and the size as well as the position of spots can be influenced by breeders but there is an element of randomness. There is still a lot that we need to learn!

Additional Information

i The gene for spotting is called MITF. It codes for microphthalmia transcription factor and is located on chromosome 19. There are two other genes that cause spotty sheep but in this primer we only look at this gene that causes the kind of spotting we normally see.

At the end of this chapter we need to soften the rule 'UNSPOTTED trumps SPOTTED' slightly. It is being suggested that this rule is only true in animals that are Aa/Aa on *Agouti-Locus* and that animals with other agouti alleles CAN show (small) signs of spotting even when they have SPOTTED only once. 'Suggested' and 'CAN' meaning: We don't know if and under which circumstances this is the case.

Skudde lamb - the white hair around the mouth tells us that this lamb will grow into a spotted grey

White isn't always white

As we have seen the pattern white means that there is no eumelanin – no base colour – anywhere. No black. No brown.

The sheep is white.

Note

White is a pattern, not a colour

!

Like the other patterns white is coded at the *Agouti-Locus* (abbreviated as Awt). This version of the gene prevents the use of eumelanin (thus making the base colour invisible to us) but *Agouti-Locus* has its own pigment: pheomelanin (tan) that comes in a slightly reddish tint.

In contrast to the pattern solid which doesn't have an 'on-switch' for the agouti gene, white does have that switch. There is functioning Agouti Signaling Protein being produced and used. It is even used in different 'patterns'. Black or brown can't be incorporated but tan can.

Those animals are not pure white – we see freckles or an orange spot on their neck and some animals look 'golden' or even red.

The intensity of tan can vary. From a hint of yellow or orange to a deep, rich red brown. And even though it is 'kind of brown' it looks different to brown eumelanin.

Golden sheep – very strong tan, Coburger Fuchs (Coburg Fox sheep)

British White, Shetland-Cheviot

Pink nose and eyelids – these sheep are white and spotted.
Friesian milksheep

The very first white sheep were probably slightly reddish or freckled. From these animals pure white sheep were bred but also sheep in a rich red, golden colour.

To get from a white sheep with tan to one that is pure white, the incorporation of all pigment needs to be stopped. Black and brown are not used due to the Awt allele itself. Stopping the incorporation of pheomelanin apparently has different mechanisms. Most British breeds of white sheep have a mechanism that results in white sheep with black noses, eyelids, and feet. The skin can be pigmented, too. This version of white is called **British white**.

Another version with pure white wool has (mostly) pink eyelids, noses and very little pigment anywhere. They are called **continental white**. These sheep also carry spotting. They are white with white spots. Spotting reduces or prevents the incorporation of pheomelanin.

Villsau (old norwegian breed) - white ram with strong tan

Many old breeds still show signs of pheomelanin. They might have freckles. Very often we see an orange spot on their neck when they are born. This usually fades to white or near white as the sheep matures. This version of white is called **Welsh mountain red** after the breed where it was first noticed.

Other versions of white show tan all-over or in a pattern that is similar to pattern GREY. One example is the Portland sheep.

In all of the sheep that do show tan this pigment is visible in 'hair' but fades in wool. Freckles on face and legs remain well visible while the wool gets much lighter.

Especially in rams of primitive breeds this tan can make for a very striking appearance! They often have a 'mane' (or 'beard' as it is sometimes called) made up of coarser fibres that can retain a strong tan colour.

How to find out what's there?

As we have seen in our very first game, some animals have no secrets (colour-wise). Brown, solid, spotted: we know all the cards.

For other sheep we might know one card. Or two, three, four, five...

How can we find out which cards a particular animal has?

It's best to start with what is known for the breed.

Which cards are in the pack?

Of course we can never be sure that 'known' is all there is. A pattern might be extremely rare or wrongly called something else. But chances are higher that we have a pattern that has been identified in the breed before. Or breeds – if we have crosses.

We can always include the highly unlikely cards later.

Test breeding

We can just watch our flock and take note of the colours we get in our lambs. That will tell us something about the cards of their parents. The more unknown (to us) cards are in our flock, the harder this gets.

So instead of just waiting and taking notes we can do test breeding for colour. That is easiest done with one animal that has six known cards. In breeds where they exist that would be the brown, solid, spotted animals. What you see is what you get. A brown, solid, spotted ram will not enter unknown cards into the game. The ewes will enter cards that are either their phenotype (like black) or the other (the hidden) card. If they play the card that matches their phenotype, we haven't learned anything new about their genotype but if the other one is passed on, we know what she has been hiding from our eyes!

Let's play another round of the game. That is very much one we played before. Brown, solid, spotted ram.

What are the possibilities with a white ewe?

She is **Awt/A_ B_/B_ S_/S_**

The underscore symbolises 'unknown'.

Her stacks of cards look like this:

She can play her card for white on the pattern stack. We know nothing more about her than we did before. (But we know more about her lamb as we have seen earlier.)

If she plays a pattern that is not WHITE, we will see it. All patterns trump SOLID. So whatever pattern card she plays: it will be visible. If she plays SOLID, we will also see it. The lamb will be solid because it has two cards for SOLID.

But if her second card is also WHITE, we will never know for sure. She can play the same card over and over again. We can't know if it is the same card that we are seeing or a different one that says the same thing. Of course chances get higher for the other card being played, the more lambs we have from her. But that's it.

Same for base colour. If we get a not-white lamb, we will see the base colour. Black or brown. The ram only has brown so one of the ewe's card will be visible. But only the one she plays. And again: she can play the same card over and over again.

On *Spotting-Locus* she also has two cards that are unknown to us. One of them will be visible (if she played anything but WHITE for *Agouti-Locus*).

- **spotted lamb: SPOTTED card from the ewe**
- **un-spotted lamb: The ewe played an UN-SPOTTED card.**

To wrap it up: If we get a white lamb, we know as much as before. If the ewe plays a pattern card that is not WHITE, we can fill in some of the unknowns in her genotype.

If for example the lamb is a brown based grey that is not spotted, this would be the lamb's cards:

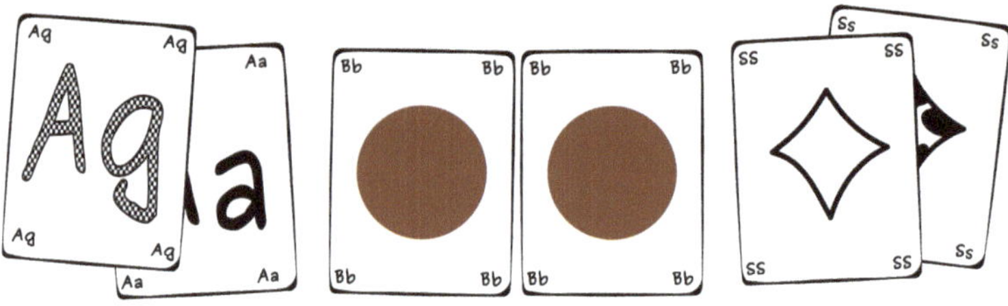

If we take those cards away that were added by the ram, we are left with Ag, Bb, SS. These were added by the ewe. She also has Awt – she's white.

Now we know more about her genotype:

Awt/Ag Bb/B_ SS/S_ Only two gaps left!

It is a game with an element of chance. She can have 20 lambs and always play the same cards. Or she can have two identical cards on one or more stacks.

Just as in a real game of cards we can't know the other person's hand for sure until they play it. We can play the game with the ewe's parents, too. Very often that reveals some information.

Let's take the same ram with a black ewe: What do we know? What are the possibilities?

She is solid and a solid animal has two cards for SOLID. She is black therefore one of her cards for base colour is BLACK. We don't know what the other one is. She is not spotted. One card for UN-SPOTTED.

Both the ewe and the ram can only play SOLID for agouti pattern. The lamb will be solid. If the ewe plays the cards that we can see on those other stacks, we won't have learned anything new about her. If she plays one or both of the other cards (and those are different to the one we see), we can fill in one or both gaps in her phenotype.

We can influence the probability for either through the ram and we'll go into more detail for planning later.

Navajo Churro

Suitable partners for test breeding

To figure out the genotype of a sheep it is best to breed it to a sheep with a fully known genotype which has two cards on each stack that are the least dominant ones in the breed i.e. the recessive ones that are trumped by all the other possibilities.

We can only use a solid, brown, spotted ram if they exist in our breed. If a card doesn't exist in our breed, we can just ignore it, remove it from the list of possible cards, and play with the rest. Sometimes we need to add them in again later because one of our sheep had a card hidden up their sleeve. But for starters we can play with what has been seen before (even rarely).

We can use another ram with fully known genotype. Known by descent or by the lambs he produced. But we are influencing our chances. If we take a solid, brown, un-spotted ram for example who also has SPOTTED, we only get half the chance of a SPOTTED card being played by him. And only then do we have a chance of figuring out if a ewe has an (invisible) card for SPOTTED.

Let's look at that stack in more detail. *Spotting-Locus.* What are the possibilities with a ewe that is un-spotted and whose genotype we want to figure out? She has a card for UN-SPOTTED. That is her phenotype and that card trumps SPOTTED. With our ram, who has SPOTTED as well as UN-SPOTTED, we have the following possibilities:

- **Ram plays SPOTTED, ewe plays UN-SPOTTED = Ss/SS**
- **Ram plays UN-SPOTTED, ewe plays SPOTTED = SS/Ss**
- **Ram plays SPOTTED, ewe plays SPOTTED = Ss/Ss**
- **Ram plays UN-SPOTTED, ewe plays UN-SPOTTED = SS/SS**

We have three (theoretical) lambs that are un-spotted (2x SS/Ss, 1x SS/SS) and only one that is spotted (Ss/Ss).

Whether those un-spotted lambs have two cards for un-spotted or just one and one for spotted we can't know because both possibilities look the same.

Even if the ewe has a card for SPOTTED, the ram can mask it with his UN-SPOTTED. We can't see the genotype of an un-spotted lamb and we can't know if the ewe or the ram or both played UN-SPOTTED.

To coax a SPOTTED card from an un-spotted ewe (if she has it) we need luck. Or lots of lambs. We'll only see it if both play SPOTTED. Chances are only 25%. With a spotted ram chances would be 50% because the ram will always play SPOTTED – the only version he has.

In planning test breedings we need to think about what we want to find out and why. Maybe I want to find out which ewe in my flock carries spotting because I don't like it and want to eliminate it from my flock. A spotted ram will reveal some of my ewes as having a card for SPOTTED. That is good to know but I'll have a whole generation of lambs that all carry at least one copy of SPOTTED.

That doesn't sound like an advancement towards my breeding goals but the knowledge about genotypes can be very valuable in the future and well worth it.

Are you just curious without having preferences? Easy! You can do whatever test breeding you fancy. For eliminating alleles in your flock you have to be a little more careful. Do you need ewe lambs as replacements for your flock? Is your first priority to grow your flock without buying (many) animals? Perhaps you shouldn't do test-breeding with an unwanted allele just now.

Focus on high quality sheep. Choose a ram that is likely to add the colour you want. Keep the task of eliminating the unwanted alleles for later.

What if it's all a mystery?

Sometimes we just can't assign a card to a phenotype if it's a pattern we don't recognise, base colour that is hard to make out, or a heavily spotted animal with a single, small brownish patch. Is it faded black? Or brown? Is there a pattern or is the animal solid?

Does the animal have two different agouti patterns that result in a confusing phenotype? Or is this ONE pattern allele? If it is a combination of two patterns: Which patterns are they?

This animal can only pass on one of its cards. If I breed it to a solid sheep, I will know one of the cards of my mystery sheep. I might see a clear pattern in the lamb that is part of the pattern mix of the parent. Half the mystery solved! Or rather more than half: I now know that the mystery sheep has a combination of two patterns. If I mentally deduct the identified ingredient I might even be able to make out the second one. At least have an educated guess.

A mysterious colour that - after pedigree analysis and test breeding turned out to be an agouti pattern

If the lamb looks the same as the mystery parent, I now know that I'm looking at ONE pattern allele that causes this phenotype. If I get a solid lamb, I can also deduce that there is one pattern allele in the mystery sheep that causes the phenotype. It is A_/Aa and the lamb is Aa/Aa.

When trying to ascertain if lamb and mystery parent have the same pattern, we need to take base colour (and spotting) into account. Patterns can seem to look a little different on black and brown. Certainly the sheep are 'not the same' at first glance. That can be misleading.

Pattern and base colour are coded on different genes. The cards are played independently of each other. So we can get any pattern in any base colour. And two black animals can throw a brown lamb.

Different base colours can make it harder to see differences and similarities in patterns. Also some patterns make it harder to determine base colour. A badgerface lamb looks pretty reddish/brownish. Are those badger stripes black or dark brown?

The colour of the eye rims can be a giveaway: They are usually lighter coloured in lambs with base colour brown whereas in lambs with base colour black they show dark pigment.

Adult grey sheep often have a lot of tan that gives them a reddish or brownish hue but they are still black sheep. Not brown.

Note

To determine the pattern of a sheep try not to be influenced by its base colour.

!

Jacob and his sheep

Breeding with a goal: Those in the know have a clear advantage!

The earliest mention of a very clever breeder of sheep and goats can be found in the Bible. In the book of Genesis we read about Jacob. He looked after the flock of his father-in-law and one day he wanted something in return for his work. He said he will continue to tend to the flock but he asks for the spotted and black lambs for himself. Apparently there were just a few of those in the flock. The father-in-law takes away all those spotted and black sheep before agreeing. Happy that he'll get the work done for free.

But Jacob knows more! He takes some branches of poplar, almond and sycamore trees and partly strips the bark off. These branches (that now have two colours) he places in front of the sheep as they are bred. The sheep - seeing something spotted during conception - give birth to lambs in the colour Jacob desires. Spotted lambs.

The sheep of Jacob - Jacob sheep.

But Jacob is even cleverer than that: He places the branches in such a way that only the biggest and best animals see them. Thus his flock not only grows in number, he also gets the best and biggest animals!

This is a fascinating story for different reasons. He knew something. He could change the colour of the flock over generations. And the fact that it is mentioned in the Bible lets us assume that this was an amazing skill! Worth being written down.

Left: Jacob sheep

Breeding with a goal: breeding programs and planning

A word of caution before we get into planning: This book is about colour and colour only. There is much more to a sheep than its colour! Our breeding program should aim at getting our flock as close to our image of a 'perfect sheep' as we can. It is relatively easy to breed for (or against) a certain colour. Breeding a good sheep, typical for its breed with all its characteristics, is much harder!

But colour kind of jumps out at us. A lot of our first impression is made by colour. We tend to think of a sheep in a desirable colour as being better than one in an undesirable colour and we might base our decisions mainly on colour.

This book aims at seeing colour in sheep in a relaxed way. The more I know about the genotypes of my sheep, the better I can plan. The great sheep with an unwanted colour can improve my flock in many ways. Colour is relatively easy to fix in future generations.

In breeding we think in generations. In breeding sheep we also think in flocks. We want as many desirable traits in our flock and in as many sheep as we can. If all of those traits come together in one individual sheep, we'll have bred the sheep of our dreams.

Don't lose great traits because of an undesirable colour!

But also don't lose a rare pattern because the sheep is not perfect or because you don't like that colour. Other breeders might be happy to preserve it in the breed.

Left: Skudde lamb

Breed standards

A breed standard describes the ideal sheep of this breed. A goal. It is not an exhaustive inventory of what is present in the breed. If it were there'd be no need to breed. We'd just need to multiply the sheep we have because 'nothing exists that's not in the standard'. There'd be no need to constantly learn and improve our skills as breeders.

Humans have kept and bred sheep for thousands of years. Standards only came up in the 19th century. The goal was improving breeds which means changing them and also creating new breeds! Standards for old or heritage breeds mostly came up much later. In the 20th century. Aimed at preserving the breeds that were getting rare. But even those were mostly written in the spirit of usefulness. Not as accounts of all the diversity but as a vision of a perfect sheep of that particular breed. Sometimes personal tastes of those writing the standard played a part. As well as their vision of the breed 'in times gone by'.

Sometimes you hear that a certain colour that pops up in a breed is a sign of cross breeding some generations back. Often you also get a name of the breed that must have been used. There can be truth in that. But this is often claimed in breeds that don't usually have many different agouti patterns. People are not used to seeing them and have very little experience in differentiating between patterns. Sometimes the breed that is named as being the source of the colour doesn't have this particular colour! We can't even see it in historic pictures.

It can be fun to speculate about the history of 'our' breed and its relationship to other breeds based on the colour.

But no matter where a colour came from: it can suddenly appear. Whether it is written in the breed standard or not. This colour can be an undesirable trait or a wonderful addition.

Achieving our goals

How can we use our knowledge to reach our goals in terms of colour? In most cases our goals will be in one of two groups:

This I want! or **This I don't want!**

It gets more complicated if we want something that is extremely rare. We'll look at that separately later.

This I want

If what you want is a colour where genotype = phenotype, you're lucky!

Every animal that looks like your goal will only pass that on. It needs two brown cards to be brown. If you see brown you will get brown! (From this one. The other sheep involved might play another card.)

If all your animals are solid, brown, and spotted, you can expect only solid, brown, spotted lambs. (But as always in biology, you are never completely safe from surprises.)

What if you have a brown, spotted ram but all the ewes are white or black or brown without spotting? And you want your whole flock to be spotted brown? You'll get there faster if you have a ram in the desired colour than if you just have one single ewe that is spotted and brown. But the principle is the same. So let's assume it is the ram that has what we want and we have a black un-spotted ewe and a white one to breed him to. (We'll leave the brown un-spotted ewe out because she's halfway there already.)

Solid, black, un-spotted phenotype: What are the possible genotypes? This sheep can be homozygous or heterozygous. (Have two of the same cards or two different ones).

Pattern

We know she is homozygous for pattern. Solid is only visible if both cards are SOLID.

Base colour

BLACK trumps BROWN. All it takes is one BLACK card. So she can be either homozygous or heterozygous.

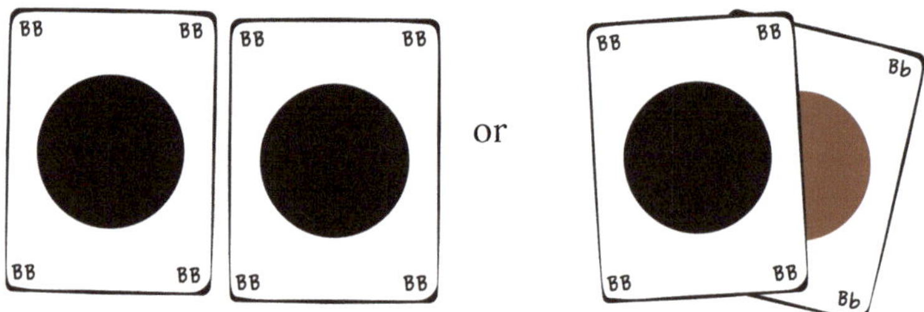

Spotting

UN-SPOTTED trumps SPOTTED. There are two possibilities here as well.

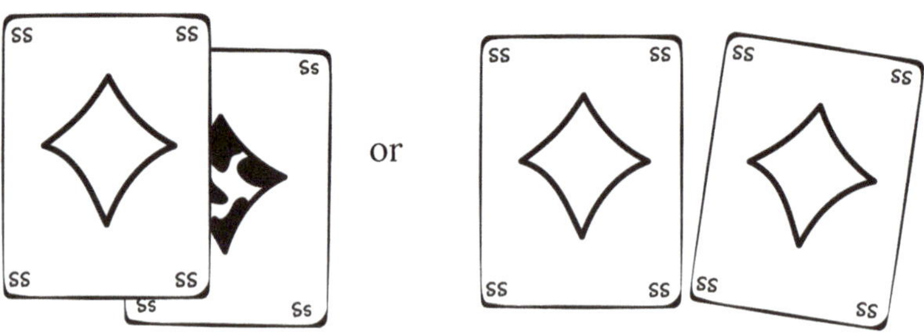

We know her cards for pattern. But we don't know which of the two possibilities she has on the other stacks. She can only play a card she actually has. Depending on the lamb she gets we can draw conclusions as to what the cards are. Which of the two possibilities she has.

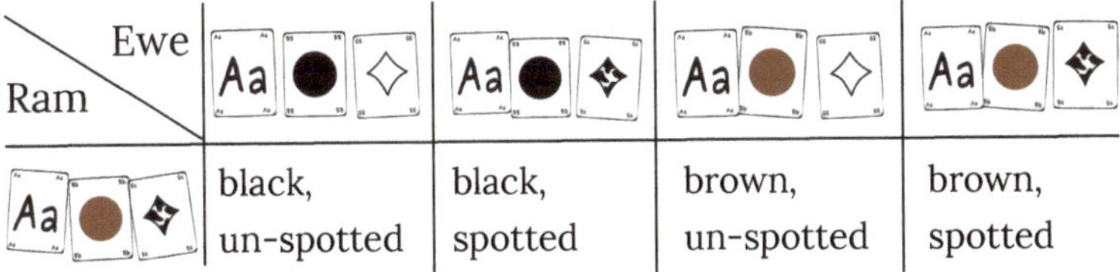

Ewe / Ram	Aa ● ◇	Aa ● ✦	Aa ● ◇	Aa ● ✦
Aa ● ✦	black, un-spotted	black, spotted	brown, un-spotted	brown, spotted

We have played this a couple of times already: find out what is there.

But this time we play with another aim: we want to find out if she possesses a trait that we want to see more of in our flock.

If we get a black un-spotted lamb, we know as much as we did before. But with a brown spotted lamb (on the right of the table) we know that the ewe has a card for brown and one for spotted. The lamb matches our goal and we know that this ewe is very valuable in regard to our favorite colour.

A brown un-spotted lamb reveals that the ewe has a card for BROWN. And one for UN-SPOTTED. We knew about the UN-SPOTTED by her phenotype and we still don't know if she doesn't have SPOTTED as well. In case she does, it was a fifty-fifty chance that she played UN-SPOTTED. For each lamb she starts with two cards and we'll get the same chance. She can produce un-spotted lambs with this ram for years in a row but still have a card for SPOTTED. Just not playing it. That would be very unlucky for our breeding goals but is no proof that she doesn't have it. It's just less likely. Same with black lambs: she can still have BROWN.

We can't know. But what we DO know: All her lambs – the un-spotted black lambs included – will have one copy of BROWN and one for SPOTTED that were passed on by the ram. That's a big step towards our breeding goals! One generation of lambs that are at least halfway there!

Breeding THEM to a brown, spotted ram will result in half the lambs being brown or spotted. Some will be both. Again: We know that they are all carrying at least one card for brown and one for spotting. And we can see which lambs have two cards.

There is no short-cut. It is a game of chance and we have to keep our fingers crossed that it works out in our favour. Knowing that the sheep all have a certain card doesn't mean they are going to play it.

Just as assuming they don't have an allele (because none of their ancestors showed it for generations back) will not prevent them from playing it.

Likelihood and chance. They are part of the fun.

Skudde lambs

This I don't want!

Removing a colour from my flock is very easy if the colour in question is caused by dominant alleles. If I'm prepared to sell all sheep that display this colour.

Say I have only black and brown sheep but don't want black any longer. I sell all my black sheep. Black trumps brown. An all brown flock means that there is no card for black around. If I only buy in brown animals, I won't get any black back in.

Note

Selection against a dominant allele is simple and quick if we select consistently. But: What's lost, is lost! We can only get it back by buying animals with that allele.

As easy as it is in terms of reaching our goals colour-wise: We probably don't want to sell all our black sheep. Some of them are likely excellent animals that we want to keep. In that case our breeding program will take a little longer. If my ewes have two black cards each they will only pass on black. But the ram will add to that one of his brown cards. The lambs will all look black but they all have one BROWN card. In the next generation – if we use a brown ram again – we get a fifty-fifty chance of brown lambs. In that generation too, we should take a good look at the lambs. The black ones might be great animals for all sorts of reasons. And they, too have one BROWN card.

A slightly more complicated scenario: A flock with white, grey (pattern grey on base colour black), and solid black un-spotted sheep. I don't want grey.

Just breeding solid black to solid black is easy: pattern grey is dominant to solid. If I get solid as a phenotype I can't have grey. It gets trickier with the white sheep. I don't know what else they have. If I keep breeding white to white chances are lower for them to show the other card. Either ram or ewe might be homozygous for white. And even if they are heterozygous, just one of them needs to play a white card to trump whatever else is there.

In breeding the white sheep to a solid black sheep chances are higher that I can see some of the hidden cards of my white sheep. If that hidden card is GREY I will have another grey sheep that I don't really want but I gained valuable information about one of the whites that helps me in my planning:

I can stop breeding the white animal that also has grey. Or I can breed it to a white in the future. Chances are high that I'll get white lambs. If bred with a homozygous white sheep ALL the lambs will be white. Phenotypically. The GREY card can still be played. Like any other pattern allele it can be passed on for generations.

Skudde rams

I want that rare colour!

What if I want to add a certain colour to my flock? I need at least one animal that has the card (or cards) that make up this colour that I don't have in my flock, yet. It is best to get the desired genetics in a ram because half of the genetics of all lambs will be his.

If I'm lucky enough to find a ram with the desired colour, it will be pretty straightforward. But let's assume I can only find a single animal with that colour and that animal is a ewe. In most cases this rare colour will be a rare agouti pattern. One ewe showing this pattern. Let us assume a breed that has (as a breed) all cards that we talked about previously. Our pattern - whatever it is - will trump solid and be trumped by white. In combination with any other pattern it will appear as a mix of both patterns.

I won't choose a white ram for this ewe. The resulting lamb might have the pattern I want but chances are that it is hidden under white. I want to see if the offspring has the pattern or not. Using a solid patterned ram makes a lot of sense. It doesn't really matter if base colour is black or brown. Some patterns are easier to see on black but base colour is inherited independently so I can tackle that later if I want to. I'm looking for a solid black or brown ram. That is an exceptionally good animal! Since in this scenario I'm using a ram of a common colour, there is no need to compromise.

All that remains for now is keeping fingers crossed for a ram lamb. This year or next or the year after.

Now we need to think in generations again: if I breed all my ewes to this solid ram, I might be lucky enough to get a ram lamb in my favourite colour out of this one ewe that has that colour. But he'll be a half brother to all the other lambs. I might be better off using one ram for the ewe with the favourite pattern and another ram for all the other ewes. Maybe there is an opportunity to send her to another breeder to join his flock for breeding.

We are aiming for a ram lamb with the desired pattern that doesn't have the same sire as most of the other lambs. But we could also sell off all female lambs. Or make breeding groups with different rams (once our 'ram of a favourite colour' is old enough): the ram lamb gets to breed all unrelated ewes and another ram gets to breed his mother and half-sisters.

If we're only half lucky, we might get the desired pattern but on a ewe lamb. We can keep her and use the same strategy as with her dam. Or we can give her to another breeder who is also enthusiastic about this pattern and will try to breed for it.

Really important when breeding for a pattern that goes back to a single foundation animal: documentation! Back to this animal. If later I want to breed two animals with this pattern, it is helpful to know how far back in the pedigree the common ancestor is.

The lamb on page 74 as an adult ram. Skudde

It is relatively simple to 'spread' an agouti pattern. All it needs are unrelated sheep that have one SOLID card (better two). I might need to buy in those unrelated solids. Or I can find other breeders who are interested in this pattern and we act as a group – passing on animals with this pattern. Documentation is key!

But can I ever breed two animals with this pattern to each other? They have the same ancestor! Yes. They do. But how far back? How far back do you normally check? In case of a pattern that was passed on from a single animal we SEE the relationship. Colour is so prominent in our perception that we are aware of the shared allele(s). But there won't be more (or less) inbreeding just because we see one trace of a common ancestor.

On the other hand: is there a reason to breed those two animals to each other? Wanting an animal that is homozygous for the pattern allele is not a convincing argument in itself. It will make it easier to spread the pattern but that's about it. If they are fantastic animals that give you high hopes of even better lambs: use the same judgement you normaly use to decide on breeding related animals!

Shetland sheep, spotted, black and tan (gulmoget), base colour black

Hooray for diversity!

Intuition might tell us to do the following when breeding for a certain colour: get animals in this colour and breed them to other animals in this colour. Keep those that also show this colour.

This works pretty well. If my favourite colour is made up of recessive alleles (cards being trumped), all the animals will have only those cards and can only pass those on.

If my favourite colour consists of dominant alleles (at least in part), I will still have good chances of getting more and more of those cards in my flock and less of the others. Plus chances of two of the others being combined (so that they become visible) will lessen.

The homozygosity for those cards in my flock will increase. I will never know how many of the others are still around until two animals that have a recessive card are bred and both pass that card on. I might decide to sell those sheep. Maybe sell the parents too. Remove those cards from the packs of cards that symbolise my flock's gene pool (all of their cards combined).

This is a strategy that will work quite well for our goal (just this one, colour) but it neglects one of the principles of nature: hooray for diversity!

To understand this fundamental principle – nature's striving for diversity – we need to go back to the full pack of cards each sheep has: 54 cards. Each card symbolising a chromosome. All that is written on the same card will be passed on together. We just look at colour genetics in this book but there is much, much more. And a lot of it is way more important than colour.

Left: Sheep on North Ronaldsay, Orkney

There is a term in genetics that is 'linkage disequilibrium'. It describes the extent to which two traits are inherited independently or together. Traits on different cards will be passed on independently. Each card on each stack can be played regardless of what happens on the other stacks. But two traits – or rather: the versions of two traits – that are on the same card will be passed on together. Our ram might pass on one particular version of a trait and one particular version of another trait. They could be on the same card. Or they could be on different cards and it is pure chance that he always plays those two cards. Linkage disequilibrium takes that into account. With enough data we can calculate if those two traits are passed on together or independently. Same card or different cards. Coupled or pure chance!

Pure chance: in a breed that has black and brown base colour and un-spotted and spotted we can see that they are inherited independently. Spotted and brown are both trumped by other cards so that we see them less often but the ratio of spotted to un-spotted remains the same. It is pure chance – not a coupling of genes. As always in statistics we need a sufficient number of 'cases' to make it reliable.

What if we find that there IS linkage disequilibrium? That it is not pure chance?

A chromosome is passed on as a whole. Not just bits of it. We are not playing with half a card. If genes A and B are on the same chromosome, the versions that are coded for those genes will always show up together.

That is actually just the simplified version. Nature found ways to turn 'always' into 'mostly'. There are ways to exchange bits of cards with bits of other cards. The same bits of the same type of card.

This 'exchange of bits of cards' is very rare but it does happen. What is important to know is that variants of a gene are almost always passed on with the variants that are on the same card. Let's make up an example in sheep: Imagine there was a single gene for staple length (length of the wool) and imagine this gene was on the same card as the gene for base colour. Then we could live in a world where all black sheep had long wool and all brown sheep had short wool.

If just once in the long history of sheep bits of the cards did get exchanged there might be a (rare) version in the genepool that has black AND short wool / brown AND long wool. Always being passed on together in THESE combinations. The other combinations (black and long, brown and short) are much more common. But they are not the only ones that exist. Calculating linkage disequilibrium will not give us 0% for one and 100% for the other. It will give us a number that will say 'those two traits are mostly coupled in this combination'.

Chances are high that the chromosomes that have information about colour also have information that is important for health, fitness, and longevity. Of course there is no single gene for 'fitness'. There are lots of genes that play a part. Some more so under certain circumstances. But one genetic aspect to fitness might be on the same card as a gene for colour. And one version of that gene might be coupled with a certain version of that colour gene. Maybe that version is helpful in fighting a virus. Or will be helpful in fighting a virus that is yet to evolve! By removing a version of a colour gene from the gene pool we might remove something as important as that. We don't know.

A lot is known about the sheep genome. There is more that we still need to learn.

With colour genetics being rather simple we can often have both: genetic diversity AND the colours of our choice. Most people have preferences. Breeders of pedigree sheep follow a standard that often calls for certain colours and excludes others. Colour is so important for our first impression of an animal that we can hardly ignore it. But we don't need to limit genetic diversity unnecessarily!

We talked about the solid, brown, spotted animal a lot. If that is the colour we want, we can't have diversity of the cards that go with it. Those animals will have two of the same cards in each stack.

But with all other colours we can aim at genetic diversity and still get the colour of our choice. We can add scientific knowledge to our intuition.

An example: our breed of sheep has the 'official' (registered) base colours black and brown and the patterns badgerface (Ab), black and tan (At), and solid (Aa) as well as SPOTTED and UN-SPOTTED.

If I favour badgerface on a black base colour I could just use black based badgerface sheep. With a good bit of luck my flock will not have any of the other cards in a couple of generations. I will only ever get animals of my favourite colour. But that is not 'diversity'. What if I used a solid, brown, spotted ram? He is not what I like in terms of colour, but: the next generation will have two different cards on each stack. And still all the animals will be badgerface, black, un-spotted. Just what I like.

For the next generation I won't use a solid, brown, spotted ram again. Since all the lambs have one of those cards each, a ram like that would raise the chances for solid, brown, spotted to fifty-fifty (for each of those independently). I could use a ram that is (likely to be) homozygous for badgerface, black, un-spotted. (We looked at how to figure out genotypes in previous chapters.) All my lambs will be badgerface, black, un-spotted. For each locus – independently of the others – chances for two different cards are fifty-fifty.

If I use a ram that is heterozygous for all or some of the genes I will have lambs that are heterozygous for some genes and homozygous for others. Those lambs lack genetic diversity for those loci but some will look the way I want. Others will lack in that respect, too.

We should aim to strike a balance here. Do I need lots of lambs in my favourite colour to keep in my flock? Go for a ram that is (likely) homozygous for the favourite colour!

Do I mainly need lambs to sell? Why not use a ram that will add diversity? I can still keep the best lambs that also happen to be in my favourite colour (they will be heterozygous) and sell the others. I can play for diversity in years I want to keep just a few and play for my favourite colour in years when I need many 'keepers'. If I breed good lambs, there will be buyers that either have no preferences for colour and also buyers that want exactly that colour.

Playing for diversity is not aiming at 'each individual sheep carrying two different cards for every stack'. It aims at keeping all sorts of versions of cards present in our flock – or in the breed in general.

If we document well over a couple of generations, it isn't extremely hard to figure out the (possible) genotypes of our sheep. With that knowledge we can keep cards in the game (our flock) and still have a flock in our favourite colour.

If brown is my favourite base colour, I won't get that as a heterozygous. But I could still think about allowing black in my flock to allow for more diversity.

Advanced patterns

The patterns we have talked about so far – Ag, Ab, At, Aeb and Albf – are not the only patterns that have been identified. There are more that have been found in classic breeding experiments or through observation. Breeding experiments can show whether a pattern breeds true: Whether it is passed on through the generations or whether it is a combination of different aspects that separate in breeding.

Patterns are umbrella terms. How exactly a pattern looks on any individual animal can vary. One animal shows the pattern very dark or distinctly. Another one has wool or hair that bleaches a lot in sunlight. Another one turns (age) grey at a young age. Combining information of different related animals helps in finding out what is on that pattern card and what is modified in an individual.

In theory we can get new patterns. While DNA is copied, errors can happen. Despite there being proofreading mechanisms. Bits can get lost or replaced and there can be recombinations with other stretches of DNA.

The large number of patterns we see today probably all originated from one pattern. That of the wild ancestor of domestic sheep. We have no reason to believe that this diversification has come to an end. But there is also no reason to believe that this happens all the time. And in our flock. If we see something new, it is rather more likely that we see a pattern that is rare in our breed or one that nobody noticed before.

Some patterns are hard to recognise. Gotland sheep for example are said to have different pattern grey variants.

Left: Shetland lamb, gulmoget (At)

If our sheep show patterns we don't recognise, it is a good idea to talk to breed societies and other breeders. If we have mixed breed sheep, we can have a look at the breeds that are (or might be) part of the mix. In mixed breed sheep we sometimes get a separation of alleles that (nearly) always go together in the founding breed. Or the other way round: A combination that doesn't exist in the founding breeds. This can help in learning about the colours of those breeds.

When talking to other interested people, keep in mind that some terms could be used to mean different things depending on whom you are talking to. 'Grey' is often used for any sheep that has 'kind of grey-ish wool'. It doesn't necessarily mean we are talking about pattern grey (Ag). 'Brown' can be used to describe that overall impression of a sun bleached fleece or that of a sheep with lots of pheomelanin (like a badgerface).

Also most breeds have their own terms for certain patterns. Often these are historical and a nice bit of 'sheep lore' and cultural heritage. A breeder of Shetland sheep might talk about katmoget, someone in Iceland will call that pattern golsótt, and a breeder of Dutch Drenthe Heath sheep will call it dassenkopp.

As with anything that is written in a standard: The terms used are the ones that were around when that standard was written and that conformed to the breeding goal that was set up at the time. They are never an extensive list of what exists or existed. Rare patterns might have been overlooked or deemed irrelevant due to their rareness or excluded on purpose to achieve a more uniform breed.

As to what SHOULD be in a breed: That is up to breeders and breed society members. This book wants to explain the basics of colour genetics so that breeders can make informed choices. Whichever direction they want to go.

hair sheep

Bovec sheep and Gotlands

Drenthe Heath sheep

Shetland sheep

There's more: other stacks of cards!

So far we have played with three stacks of cards out of 27. The three stacks that have known variants for 'colour genes'. Some of the others also have information that influences the overall colour impression.

Some are known in their function (or at least their effect), others suspected, and there is lots we still need to learn! We might see the effect but not know what causes it and how it is inherited.

In dogs and horses (and mice, of course) we know a lot more about colour genetics. In sheep there are not as many scientific studies. That is not due to technical reasons but due to the market. A lot is known about genes and genetic markers in sheep but most of the work is aimed at traits like slaughter weight and meat to bone ratio. Traits that an industry is interested in. Therefore we have scientific publications looking at white in merinos. But very little recent publications (making use of recent technology) that look at agouti patterns in old breeds.

The biotech company Illumina has a 50k sheep chip that has 50,000 markers for traits that are important for the industry. The lovely colours of heritage breeds don't attract huge research funds. But work is being done and we will learn more!

Until then it is up to breeders to come up with hypotheses to run against pedigrees. In our game of cards at the end of the book we included some joker cards to do just that. They can stand for anything we can think of and using them can help to shed some light: Is there test breeding we could do to help with that?

How could we see if traits are coupled? (E.g. on the same card as Agouti patterns? Are they always passed on together or can they be combined freely?)

Say that our black ram's wool gets sun bleached to a shade of brown. Same with his offspring. Is there something hereditary that causes it? A stack for light fastness (fading/non-fading)? Or could that information be included on his card for pattern or base colour? Watching his offspring over the generations, we might see the fading separated from his base colour and/or his pattern. Or we might always see it in the same combination.

There can be lots of other causes that lead us on the wrong track: It might be due to the structure of his wool fibres. But THAT might be coupled with colour cards as well...

Some genes are known in sheep. Some are known to be on a different card than the colour genes we looked at so far. Even if those versions of cards – those alleles – have not been reported in our breed we can still play with those cards and see if it helps to solve our puzzle.

The rules are the same for all breeds of sheep. Genetics doesn't work differently in one breed or another. But some cards are common in one breed and have never been seen in another.

Extension-Locus

One of the easier and better known additional stacks is *Extension-Locus*. Or *E-Locus* for short. It codes for melanocortin-1-receptor. Which is also involved in red haired humans and blonde Labrador dogs. In sheep its other name is a give-away as to the effect it causes: **dominant black**.

We have learned that we only get all black sheep (or brown – an all-base-colour sheep) if we have two cards for SOLID on our pattern stack. All other variants make Agouti Signaling Protein and are white or patterned.

If the dominant black card is involved, the effect is as if it turns off Agouti. Whatever the agouti / pattern cards say: It doesn't happen. Base colour everywhere.

Spotting is not affected. Neither is base colour.

Dominant black is another confusing term. It can show as either black or brown, depending on the version of eumelanin coded for at the *Brown-Locus*.

The abbreviation for the allele is **ED** (*Extension-Locus*, dominant black). The other variant is abbreviated **E+**. This is the wild type that allows us to see the pattern cards as usual. Our sheep needs to have E+ twice for that. ED trumps E+.

In general, ED is rare in sheep. But in some breeds it is the norm. Black Welsh Mountain sheep for example. If we cross a Black Welsh Mountain sheep with another breed, we will get black lambs. The Black Welsh Mountain passes on one card with ED to the offspring. Those lambs are heterozygous. When they have lambs they will also pass on one of their cards. If that is E+ (being combined with another E+) we will see what is in the pattern stack in the next generation.

Black Welsh Mountain

Dorper

Dominant black is the first suspicion we should have if two solid animals have lambs that are white. Or grey. Or badgerface. Or any other pattern.

Other breeds that often (or always as far as we know) have ED are Dorper and Jacob sheep. Jacobs also show that ED does not influence *Spotting-Locus*.

If we do get white lambs out of two non-white and non-solid sheep (e.g. a grey and a badgerface), we can't play ED. Dominant black masks agouti. If we see an agouti pattern, it is obviously not masked. We can conclude that there is no ED present.

Melanocortin-1-receptor is the receptor to which ASIP binds to do its job. If the receptor is changed so that ASIP can't bind, it can't do anything. ED animals are always solid!

Other animals that are well studied for colour genetics show a variety of alleles at E-locus. More than two (ED and E+) might exist in sheep as well.

Jacob

Note

Dominant black masks the pattern card. No matter what's on it. A sheep with a visible pattern like WHITE or GREY can NOT be dominant black.

'Blackish' for example could be an allele at E-locus.
A variant that causes black heads and legs and maybe 'paints on top' of other patterns. Thus masking their display on heads and legs.

'Blackish' is the colour of Hampshire Down for example. They were very popular for improving landrace sheep in continental Europe at the end of the 19th and the beginning of the 20th century. They might have left a legacy in those breeds that now causes confusing colours.

Hampshire Down - dark head, dark legs, light coloured wool

KIT

White lambs out of two coloured sheep should make us think about *E-Locus* first. But sometimes we get white lambs out of two non-white, non-solid sheep. They can't be Awt that was masked by ED. ED prevents ASIP from doing its job, resulting in solid animals. They could be extremely spotted. One large white spot covering the whole sheep.

But they might be something else. There is a gene that is well studied in horse colour genetics: KIT. In sheep this gene is located on chromosome 6 and is involved in development of melanocytes during embryogenesis. The cells that produce melanin.

This known function and a look at horse colour genetics makes it a likely candidate for effects on sheep colour.

Pigmented Head

This type of spotting shows pigment mostly on the head of the sheep. It has been described in hair sheep and fat tailed sheep.

Pedigree data suggests that this is a separate locus - not the *Spotting-Locus* we talked about so far. This gene or the chromosome it is on has not been located yet.

Different alleles are proposed for this locus. One example being the allele **Pigmented Head Persian (PhP)**. Sheep homozygous for PhP display a coloured head and a white body. The coloured area might extend down to the shoulders.

spotted lamb with ticking - which is invisible at this age

the same animal three years later

Ticking

Ticking shows as small pigmented spots on the white of a spotted sheep. When lambs are born, these areas are pure white and pigmentation increases over time. As shearlings they often still have large white areas that darken in later years.

Sometimes these animals have areas that remain pure white. Such as a white blaze on a face that otherwise gets more and more pigmented.

The allele for ticking is dominant to the one that retains white areas.

Dilute

Dilution is another one that is well studied in other animals and is proposed to exist in sheep.

In an animal that is homozygous for dilute the pigment of the base colour seems to be diluted. In reality it is not 'thinner' and there is not less of it. The effect is caused by pigment being clumped up which can be seen under a microscope.

This only affects eumelanin. Pheomelanin is unaffected. Diluted black eumelanin looks like a steelish grey. This effect is not caused by black and white hairs side by side (as in Ag for example). On brown eumelanin the overall effect is that of 'lilac' – a slightly purple looking colour. In those animals that are known to have dilute we see the wild type (not dilute) as the dominant allele.

It is just a model!

We should always remember that we are playing a game the rules of which have been figured out while playing it. And it is just a model. If it works to explain things: fine! Let's play another round! If it doesn't: start over. Play again and try to figure out the rules that apply. Find a better model!

The first three stacks we played with (pattern, base colour, spotting) are well documented. The rules have been tried and tested by scientists and breeders all over the world for many, many years.

Other theories were proved wrong, some proposed alleles (mainly agouti alleles) turned out not to exist and others were shown to be more than one allele.

There is no dishonour in that. That's just how it works. Watch, document, come up with a possible explanation or hypothesis. Then: watch and document some more. Shift your angle, look elsewhere. Refine and polish and sometimes scrap the idea.

That is the reason for this chapter: Nothing in this book is THE TRUTH! It is a model that works well to describe the truth – the real rules of the game. The further we move away from the basics the more we'll need to come up with our own models.

Left: Shetland cross lamb

Photo acknowledgements

Claudia Schulte
pages 56, 57, 62, 78, 84, 114 (Skudde)

Karólína í Hvammshlíð
pages 26, 40 (Islandic sheep)
www.islandwollprojekt.de
www.verlag-alpha-umi.de

Kirstin Piert
pagesn 76, 105 (Jacob sheep)
www.fromshepherdsown.de

Alexander Arns
page 52 (Drenthe heath sheep)
www.arns-fotografie.de

Josef Schließman
page 53 (Moufflon)

Kerstin Pillkahn
page 65 (Frisian)

M'kesha Willenbring
page 71 (Navajo Churro)
www.dotranchchurros.com

Christine Reiner
page 74 (Skudde Ralf)

Marilyn Voigt
page 89 (Bovec sheep)

Corinna Bachmann
page 99 top (hair sheep)
www.working-squad.de

Stefanie Hartmann/Anja Winar
page 99 bottom (Bovec sheep)
www.ambergau-schaeferei.de

Albert Kerssies
page 100 top (Drenthe heath sheep)
www.schaapskudderuinen.nl

Dave Anders
page 104 top (Black Welsh Mountain)

Sabrina Lauck
page 104 bottom (Dorper)
www.anglo-nubier-sonnenhof-capricornus.de

Janet Hill
page 106 (Hampshire Down)
www.hampshiredown.org.uk

Sigrid Heilmann
pages 14, 27, 31, 49, 51, 55, 88 (Skudde)
www.sigis-schafe.de

Saskia Dittgen
page 10, 19, 25, 42, 50 bottom, 96
www.havellandbordercollies.de

Irina Böhme
pages 9, 20, 22, 26 oben, 32, 34, 39, 41, 45, 48, 50 top, 52 bottom, 53, 54, 59, 64, 66, 90, 108, 110, 113
www. driftwool.blogspot.de
www.skud.de

Thank you!

This book would never have been written without all the people who generously shared information about their sheep and who didn't dispair whenever we talked about their 'solid coloured black and white sheep' or asked for pictures of a sheep's sire and dam, grandsires and -dams, and other relatives.

We hope you enjoy the book and we look forward to be playing another round of 'coloured sheep' with you in the future!

A special thanks to all the photographers for their beautiful pictures and to Jan for help with graphics; to Tracy Sayre and Kenny Somerville for helping making this a better book; to Sheila Gear for generously sharing her knowledge and for feedback and encouragement along the way; to Jenny Holden-Wilde for exchanging ideas and for taking sheep colour genetics to the lab – I can't wait to see sequence data for some of our puzzles!

Thanks a million to Roger Lundie for feedback and endless hours puzzling over Skudde pedigrees, testing theories and finding fascinating 'new stuff' in an old breed that has much more colour than I ever thought! Thanks Roger! For a fun and challenging game that I hope is far from being over!

My deepest thanks to all the crofters, shepherds, sheep judges and vets in all those faraway places for sharing your knowledge so generously and making me feel welcome in your world!

Left: Skudde lamb

Index

Select bibliography

Dawie Du Toit (Hrsg), Timeless coloured Sheep, Michael Imhof Verlag, 2014

Dawie du Toit (Hrsg), The Damara of Southern Africa, 2007

R. S. Lundie, The Agouti Locus of the Sheep, in publication

R. S. Lundie, The genetics of colour in fat-tailed sheep: a review, Trop Animal Health Prod, 2011, 43: 1245-1265

S. Adalsteinsson et al., A possible genetic interpretation of the colour variants in the fleece of the Gotland and Goth sheep, Annales de génétique et de sélection animale, 1978, 10 (3): 329-342

J.L. Han et. al, Analysis of agouti signaling protein (ASIP) gene polymorphisms and association with coat color in Tibetan sheep (Ovis aries), Genetics and molecular research, 2015, 14(1): 1200-1209

N.E Cockett, T.L. Shay, M. Smit, Analysis of the sheep genome, Physiol. Genomics, 2001, 7: 69-78

P Hoogschagen, S. Adalsteinsson, J.J. Lauvergne, A badger-face-like color variant in Texel and in Dutch sheep in the Netherlands, Annales de génétique et de sélection animale, 1978, 10 (4): 517-523

C.M. Rochus et al., ASIP and MC1R mutations causing black coat colour in five Swedish sheep breeds, Proceedings, 10th World Congress of Genetics Applied to Livestock Production,

L.J. Royo et al., Differences in the expression of the ASIP gene are involved in the recessive black coat colour pattern in sheep: evidence from the rare Xalda breed, Animal Genetics, 2008, 39: 290-293

J. Gratten et al., The genetic basis of recessive self-colour pattern in a wild sheep population, Heredity, 2010, 104: 206-214

M-H Li, T. Tiirikka, J. Kantanen, A genome-wide scan study identifies a single nucleotide substitution in ASIP associated with white versus non-white coat.colour variation in sheep (Ovis aries), Heredity, 2014, 112: 122-131

B.J. Norris, V.A. Whan, A gene duplication affecting expression of the ovine ASIP gene is responsible for white and black sheep, Genome Research, 2008, 18: 1282-1293

J.D, Parfitt, A.J. Sheppy, Hebridean Colour Genetics, adaptation of a paper presented at the 4th World Conress on Coloured Sheep held at York in July 1994

D.E. Vage et al., Mapping and Characterization of the Dominant Black Colour Locus in Sheep, Pigment Cell Research, 2003, 16: 693-697

D. Hepp et al., Identification of the *e* allele at the Extension locus (MC1R) in Brazilian Creole sheep and its role in wool color variation, Genetics and Molecular Research, 2012, 11 (3): 2997-3006

U. Gaede, Das Pommersche grauwollige Landschaf, Dissertation, 1926

H. Wang, Genome-Wide Specific Selection in Three Domestic Sheep Breeds, Plos One, 2015, 10 (6): e0128688

G. Kun, Beiträge zur Charakterisierung und Verwendung der Mischwollen von ostpreussischen Skudden und Rauhwolligen Pommerschen Landschafen, Dissertation, 1995

Z. Wang et al., Genome-Wide Association Study for Wool Production Traits in a Chines Merino Sheep Population, Plos One, 2014,9 (9): e107101

S. Nanekarani, Genetic analysis of Karakul sheep breed using microsatellite markers, African Journal of Microbiology Research, 2011, 5 (6): 703-707

J. Gratten, A Localized Negative Genetic Correlation Constrains Microevolution of Coat Color in Wild Sheep, Science, 2008, 319: 318-320

J. Gratten, Compelling evidence that a single nucleotide substitution in TYRP1 is responsible for coat-colour polymorphism in a free-living population of Soay sheep, Proceedings of the Royal Society, 2007, 274: 619-626

Want more?

If you have colour genetics puzzles that you need help with, you can drop us an email:

books@driftwool.de

And if you'd like to use the cards but don't want to cut up your book: there is a file to download and print on:

www.BunteSchafe.blogspot.com

If you enjoyed the book and want to let us know: please do! Send us an email and leave a review on amazon!

The cards

Here are some cards to play along.

Three cards with the abbreviations for the alleles on the three stacks:

base colour, pattern, spotting

Remember that there are more patterns than the ones we covered in this book. These are common patterns in many colourful breeds.

And the most important card: The question mark! For all the things we don't know (yet).

Base Colour:

BB: Black

Bb: Brown

Patterns:

Awt: White
Ag: Grey
Ab: Badgerface
At: Black and Tan
Aeb: English Blue
Albf: Light Badgerface
Aa: Solid

Spotting:

SS: Un-spotted

Ss: Spotted

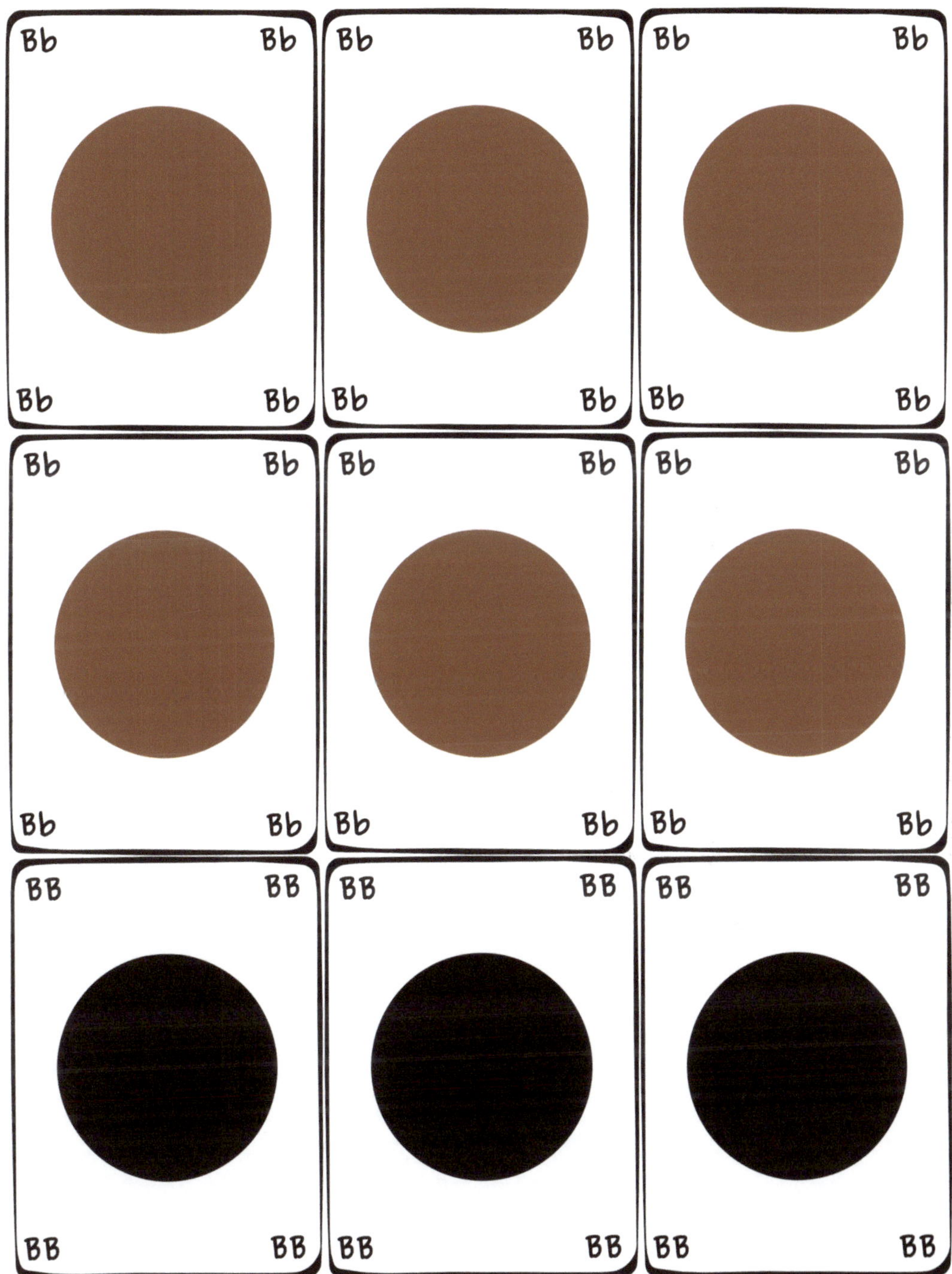
Bb Bb
Bb Bb
Bb Bb
Bb Bb
Bb Bb
Bb Bb
Bb Bb
Bb Bb
Bb Bb
Bb Bb
Bb Bb
Bb Bb
BB BB
BB BB
BB BB
BB BB
BB BB
BB BB

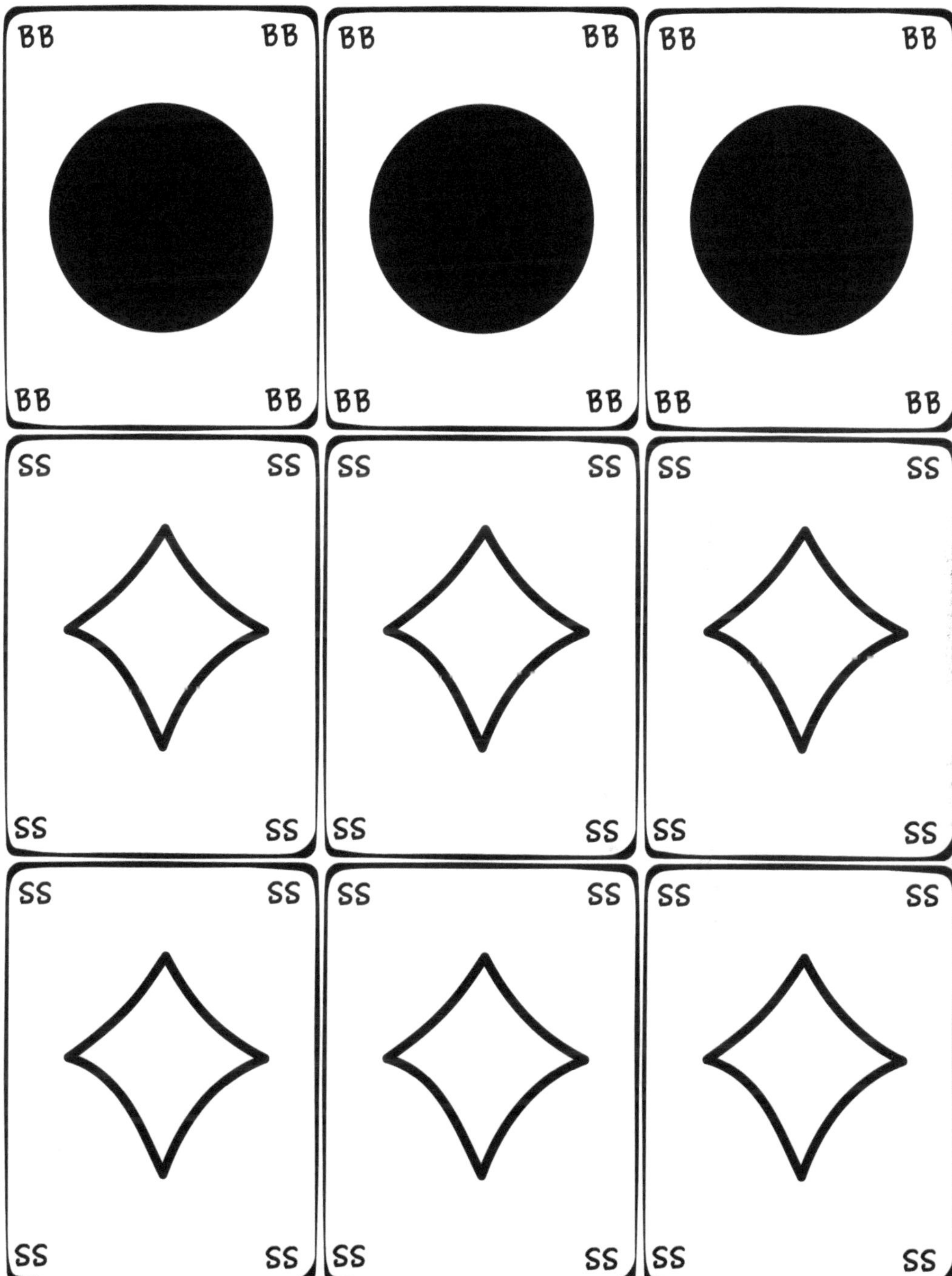
BB BB
BB BB
BB BB
BB BB
BB BB
BB BB
SS SS
SS SS
SS SS
SS SS
SS SS
SS SS
SS SS
SS SS
SS SS
SS SS
SS SS
SS SS

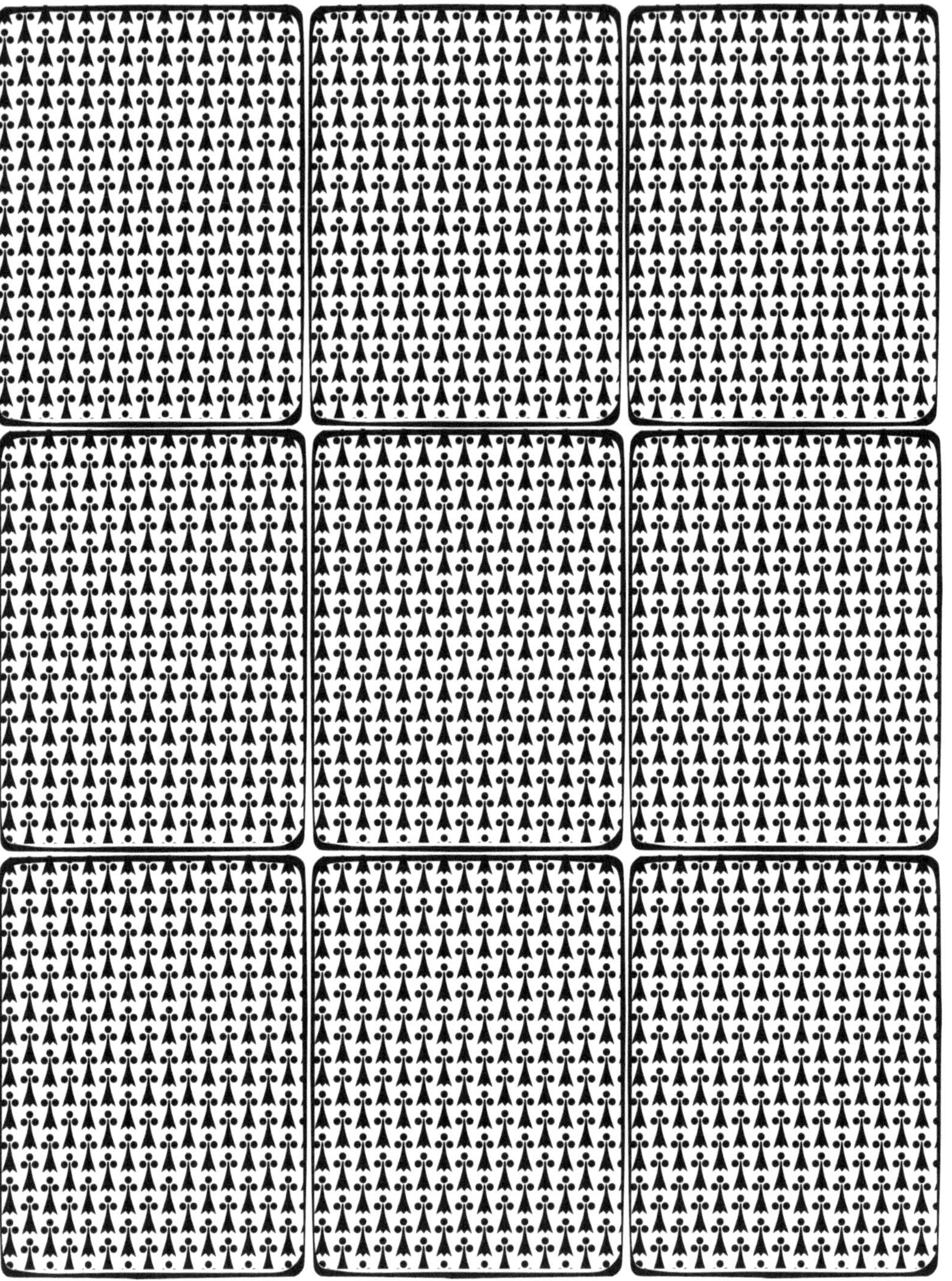

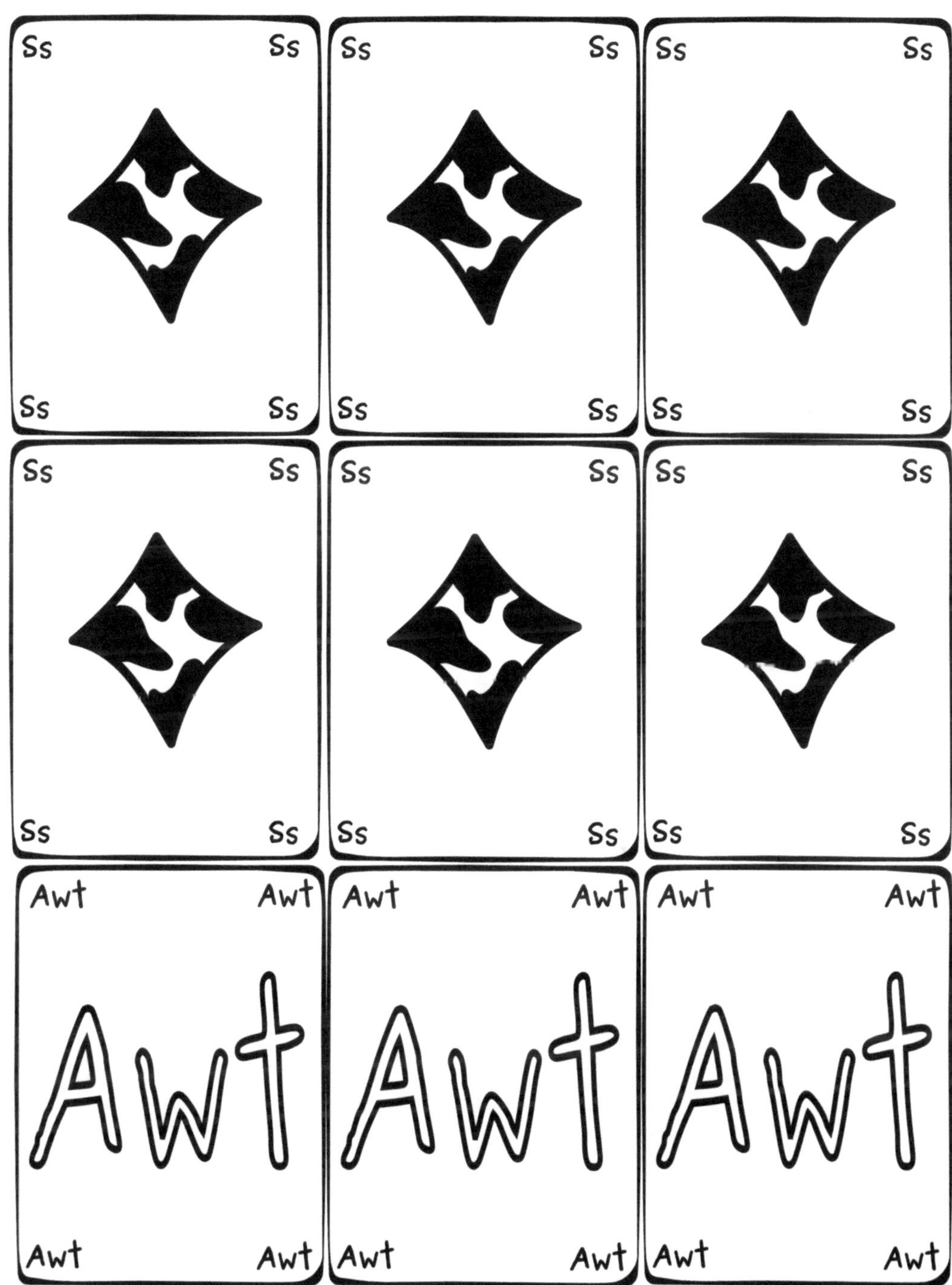
Ss Ss
Ss Ss
Ss Ss
Ss Ss
Ss Ss
Ss Ss
Ss Ss
Ss Ss
Ss Ss
Ss Ss
Ss Ss
Ss Ss
Awt Awt
Awt
Awt Awt
Awt Awt
Awt
Awt Awt
Awt Awt
Awt
Awt Awt

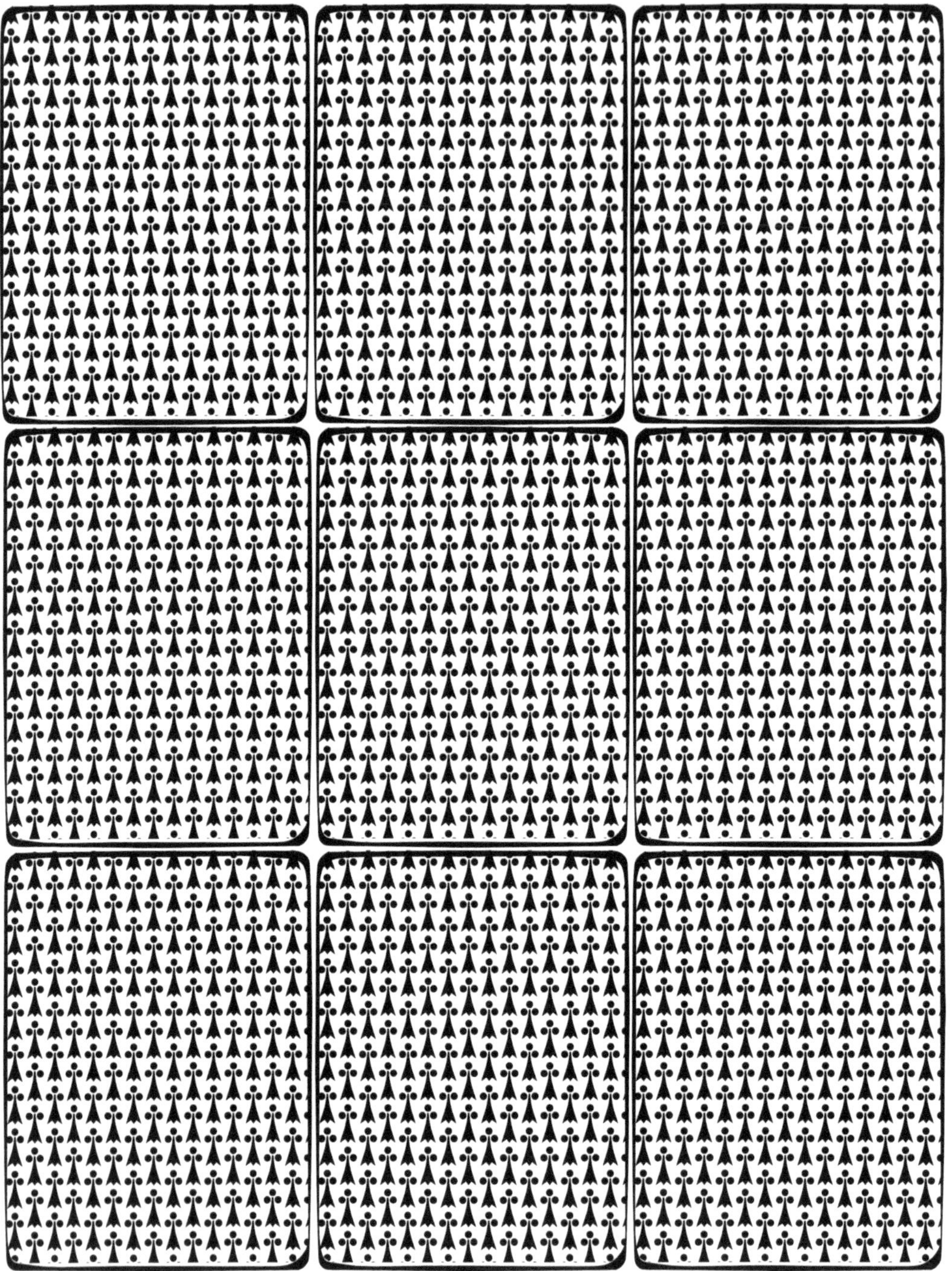

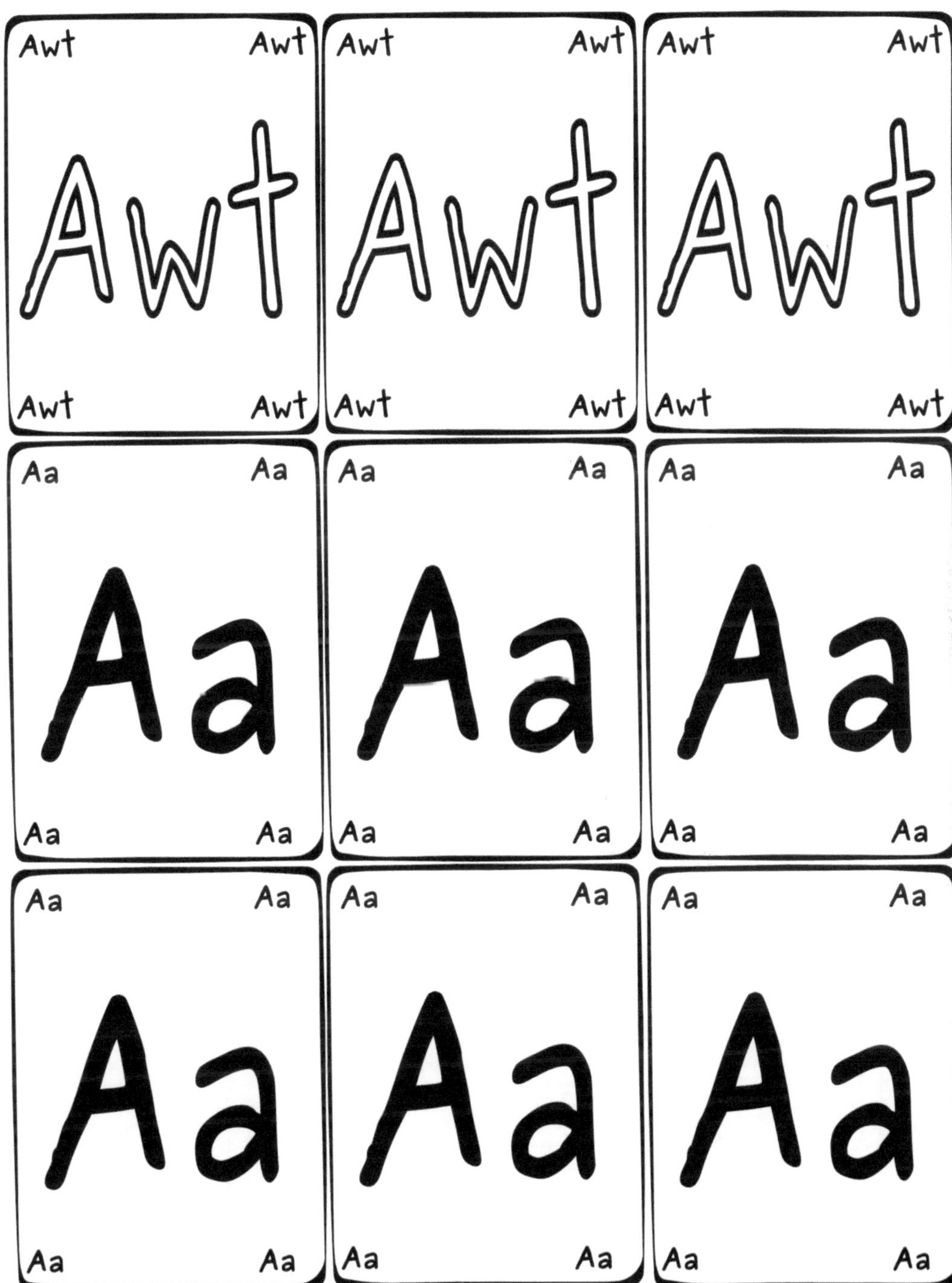
Awt
Awt
Awt
Awt
Awt
Awt
Aa
Aa
Aa
Aa
Aa
Aa
Aa
Aa
Aa
Aa
Aa
Aa

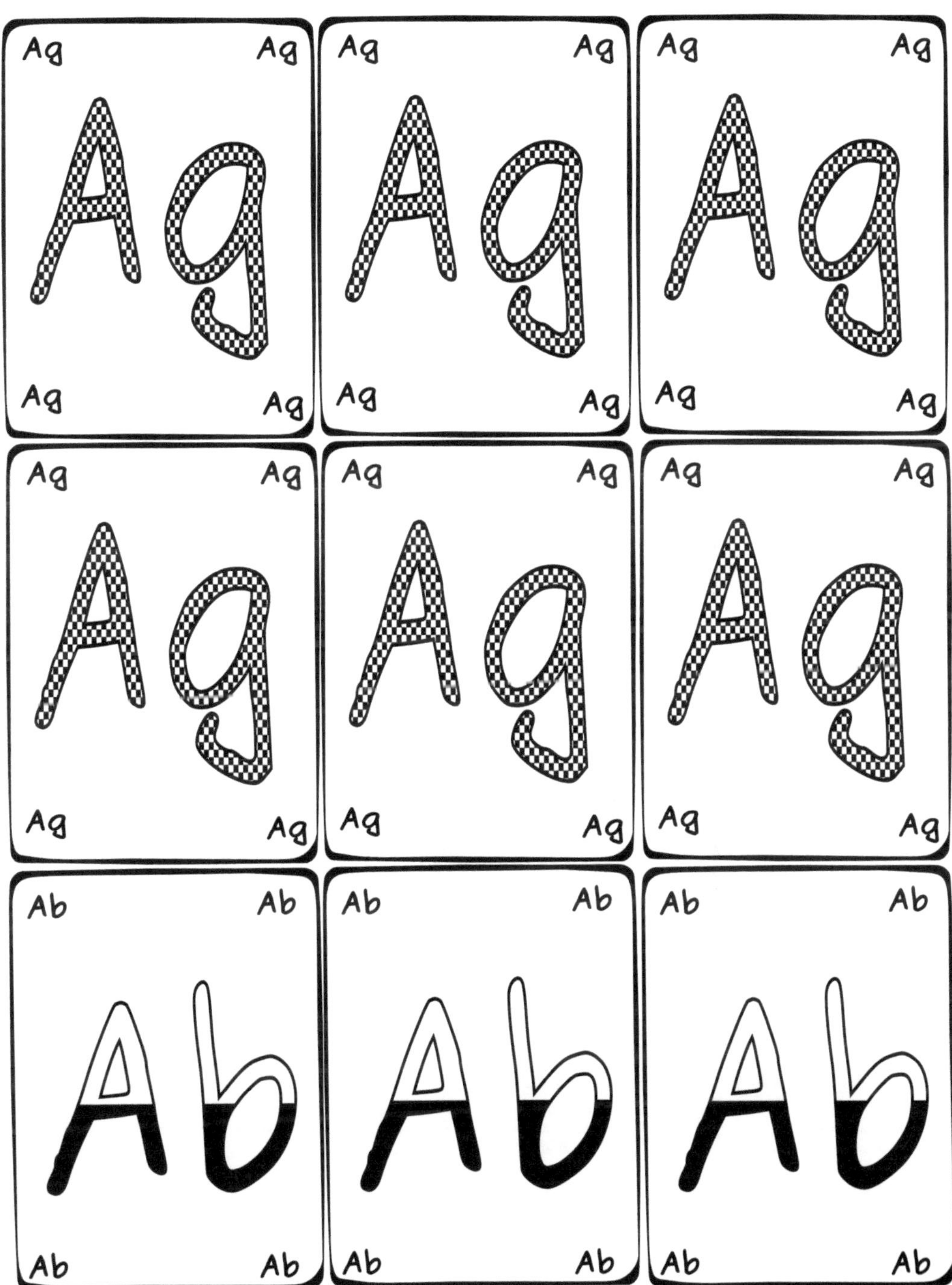
Ag
Ag
Ag
Ag
Ag
Ag
Ag
Ag
Ag
Ag
Ag
Ag
Ag
Ag
Ag
Ag
Ag
Ag
Ag
Ag
Ag
Ag
Ag
Ag
Ag
Ag
Ag
Ag
Ag
Ag
Ab
Ab
Ab
Ab
Ab
Ab
Ab
Ab
Ab
Ab
Ab
Ab
Ab
Ab
Ab

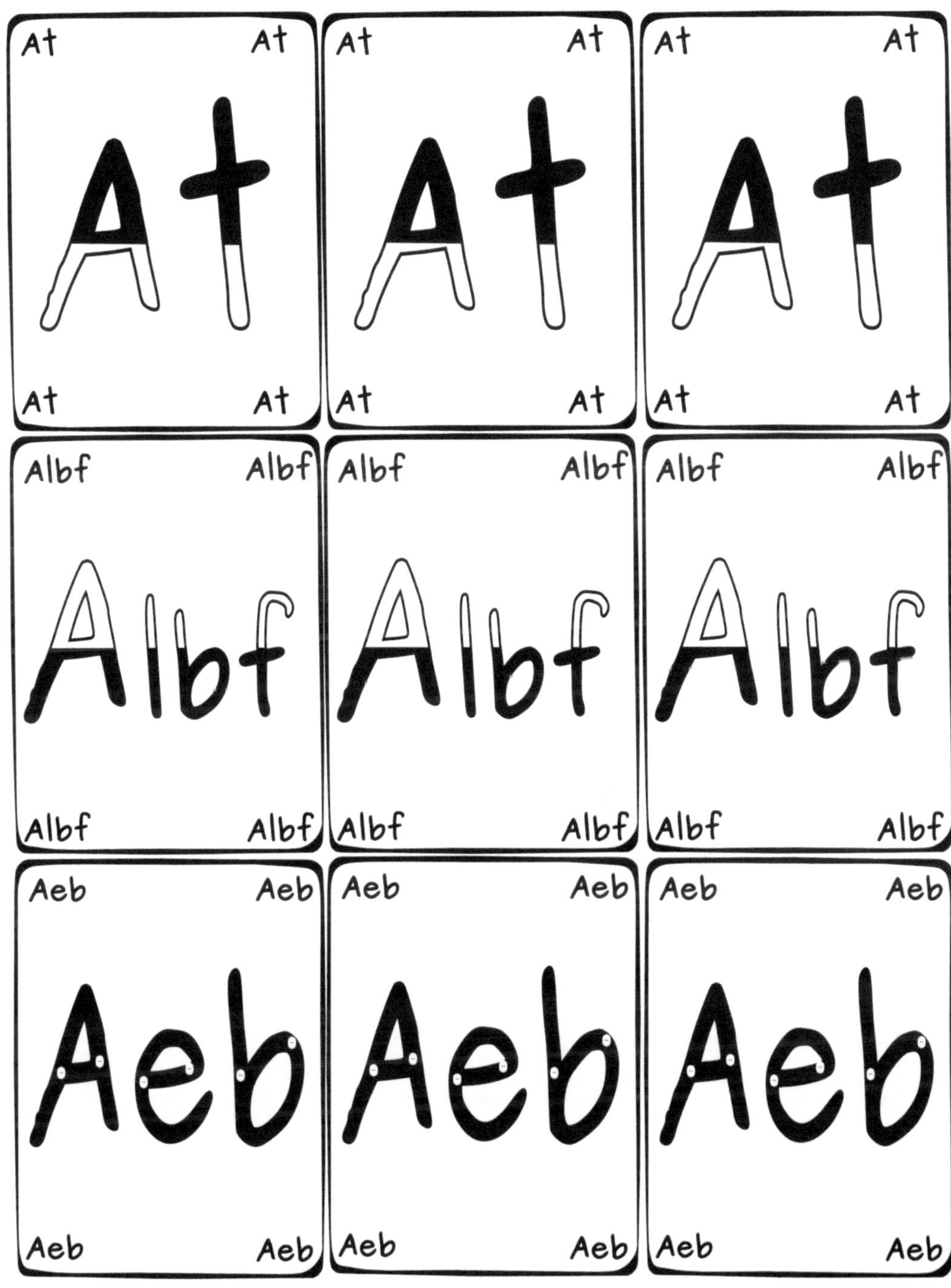
At
At
At
At
At
At
At
At
At
At
At
At
At
At
At
At
At
At
Albf
Albf
Albf
Albf
Albf
Albf
Albf
Albf
Albf
Albf
Albf
Albf
Albf
Albf
Albf
Aeb
Aeb
Aeb
Aeb
Aeb
Aeb
Aeb
Aeb
Aeb
Aeb
Aeb
Aeb
Aeb
Aeb
Aeb

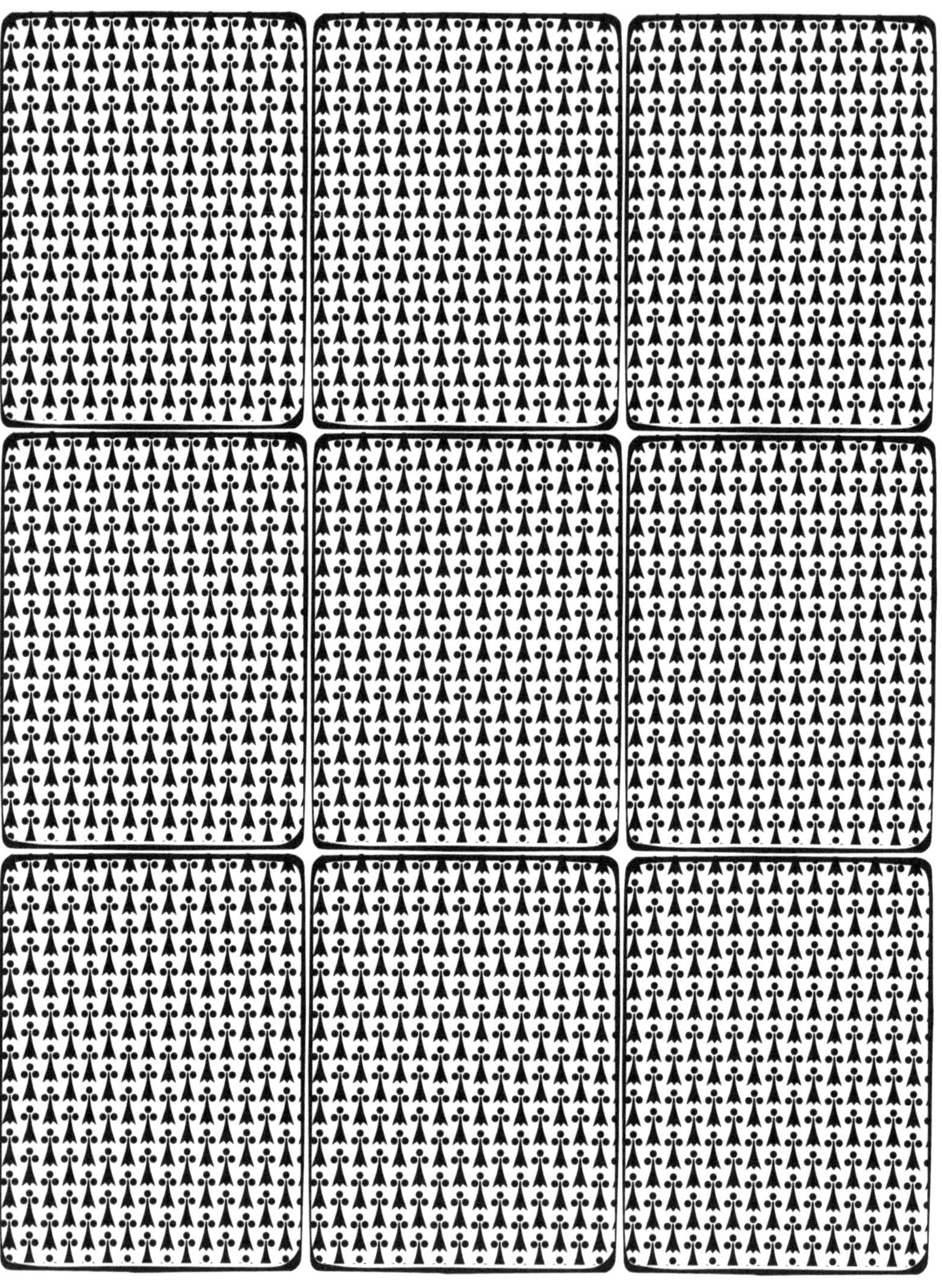

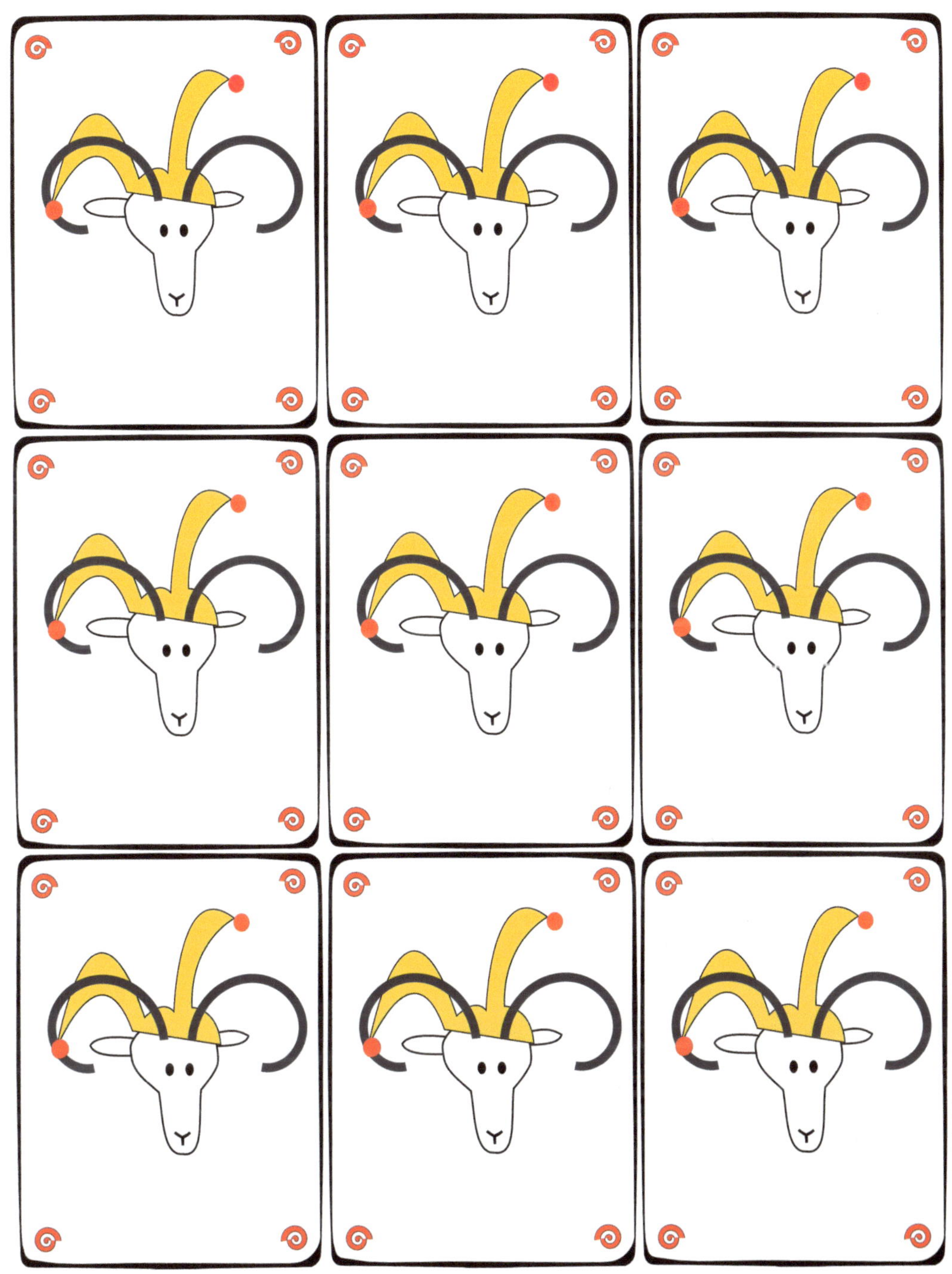

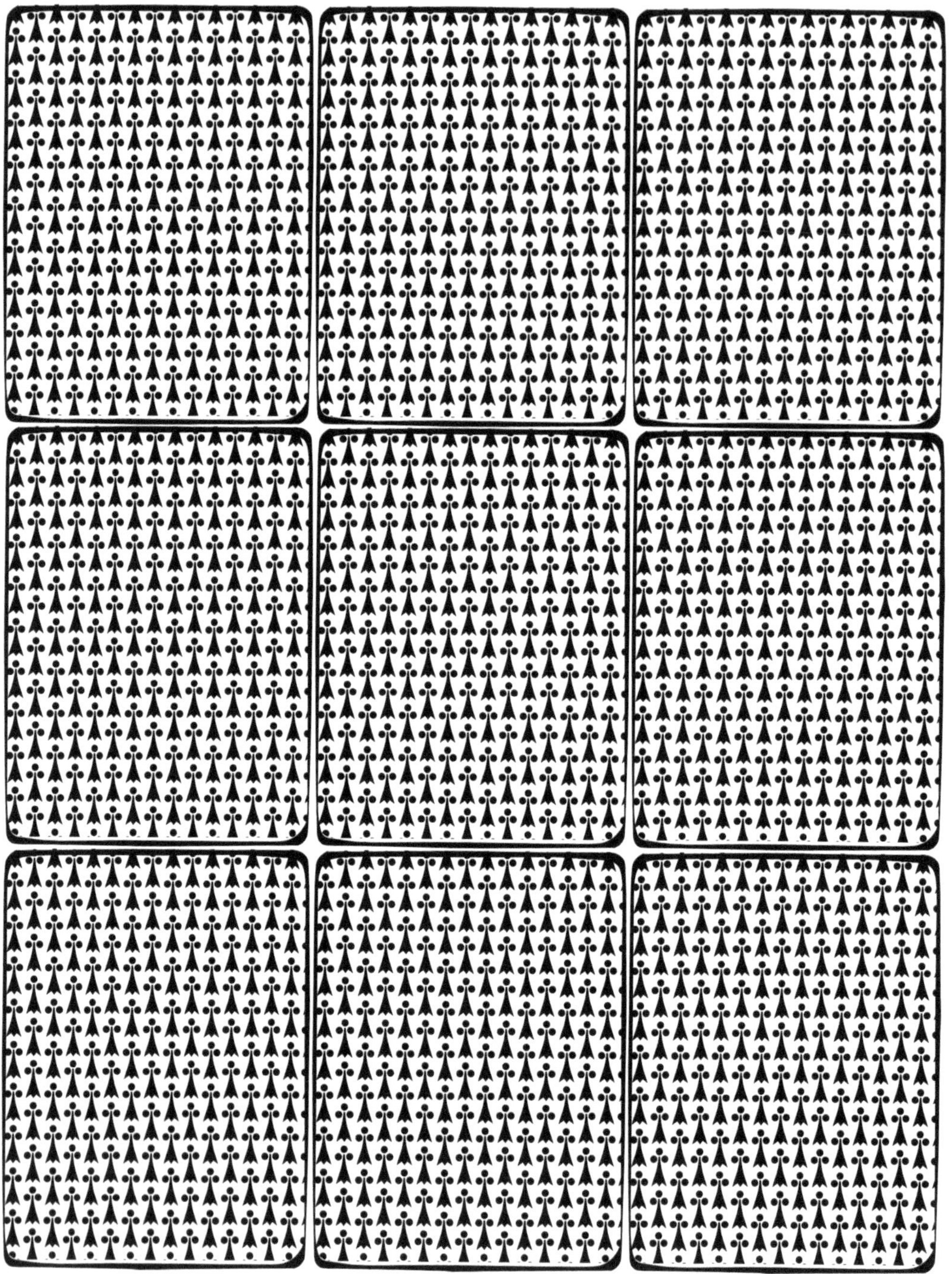

Resolution page 39

The lambs

Un-spotted grey, base colour black:

Ag/Aa BB/B_ SS/Ss

The lamb inherited grey from the ram. The ewe can only have added solid. She only has solid on her pattern stack. On the stack for base colour there must be one card for black (the lamb shows base colour black). We can't know what the second card is. Both parents could have played black. We know that the ewe could have also played brown – her second card. We don't know the ram's second card. For spotting there is one card for un-spotted (we know this because the lamb is un-spotted). That card must have been played by the ewe because the ram has two cards for spotted and he must have played one of them.

Brown spotted lamb: Aa/Aa Bb/Bb Ss/Ss

Genotype of the ram

What we already knew from his phenotype:

Ag/A_ BB/B_ Ss/Ss

We can fill the gaps:

The second lamb is spotted brown. There are two cards for solid on its pattern stack. One of those was inherited from the ram. The lamb is brown. So the ram must have played a brown card. We now know the ram's full genotype:

Ag/Aa BB/Bb Ss/Ss

www.ingramcontent.com/pod-product-compliance
Ingram Content Group UK Ltd.
Pitfield, Milton Keynes, MK11 3LW, UK
UKHW060359300726
14090UKWH00001B/31

* 9 7 8 3 9 8 2 0 7 6 1 0 2 *